Mechanisms of Resistance to Thrips and Thrips-transmitted Tospo Viruses and the Role of Crop Wild Relatives (CWR)

ATPBR
Breeding for Prosperity

ADVANCED TRAINING IN PLANT BREEDING (ATPBR)

A15, Amrut Sai Sara City, Nakshtrawadi, Paithan Road, Aurangabad, Maharashtra-431001

NIPA® GENX ELECTRONIC RESOURCES & SOLUTIONS P. LTD.

New Delhi-110 034

Dr. Aparna Tiwari
Co-Founder, Director – ATPBR

Aparna Tiwari is Co-Founder and Chief Executive Officer at the Foundation for Advanced Training in Plant breeding (ATPBR), where she leads professional training, technology liaising, awareness sessions, and other capacity building activities. She has worked previously in various positions with the FAO (Food and Agriculture Organization), Monsanto India Pvt. Ltd., ICRISAT (International Crops Research Institute for the Semi-Arid Tropics), and the MGM– Institute of Technology. She obtained her MSc and PhD from the Wageningen University, The Netherlands, and her Bachelor's (BSc) from Banaras Hindu University sity. She has authored two books and published several research papers in peer-reviewed international journals and conferences.

About ATPBR

ATPBR is a registered not-for-profit foundation, based in India, which works in close collaboration with Seed Industry, Institutions, NGOs, and Government bodies, to impart training in plant breeding to develop future-ready plant breeders. It is India's leading networking & training organization, with full-fledged expertise and resources for capacity building, market linkages, policy advocacy, and project management. Main objective of ATPBR is to develop a sustainable and thriving agricultural system to serve farmers, support emerging entrepreneurs and assist industries to build market-driven competencies in Human Resources. ATPBR is open to collaborate with diverse range of organizations, academic and research institutions, NGOs, independent professionals and aspire to meet the intellectual needs to make crop improvement an attractive and lucrative domain for all concerned.

Dr. N.K. Krishna Kumar

Former Deputy Director General (Hort. Sci.)
ICAR, New Delhi,
Former Regional Representative (Bioversity International)
South & Central Asia
Former Director, NBAIR, Bangalore

Speaker

Dr. N.K. Krishna Kumar, former Deputy Director General (Horticulture), ICAR was born on 5 September 1955 at Mysore, Karnataka State. He was awarded the Ph. D. from USA in 1993 as an East-West Scholar and PDF from UC Davis, California in 2002-03. In 38 years of his illustrious career, he served as Deputy Director General (Horticulture), Director, NBAIR, Bangalore (June 2011–August 2012), and Head (Entomology and Nematology) for 8 years. He has made valuable contributions in the field of Agriculture, Horticulture, Research management, Biodiversity in India and in the region.

Dr Kumar contributed significantly to IPM of vegetables, location of source of resistance, management of sucking pests as vectors of plant pathogens etc. As a part of Ph. D under the guidance of Dr. Diane Ullman at Hawaii under the East-West Fellowship, Dr. Kumar contributed to better understanding and management of vector thrips of Tomato spotted wilt virus (TSWV). He continued this work for his post-doctoral studies focusing on the role of TSWV glycoprotein in vector acquisition, pattern of transmission in vector population & relationship between virus replication and transmission efficiency.

Dr Kumar is a recognized advisor for post graduate research work at UAS, Bengaluru and Kuvempu University, Shivamogga. His students have made valuable contribution on molecular systematics of aphids, whitefly, thrips, role of aphid honeydew, resistance to pesticides in relation to virus transmission etc.

Dr. Kumar organized several scientific meetings such as National Meeting on Entomology and IPM, sap sucking insects, focused meetings on leafhoppers, whiteflies, mealybugs, insecticide resistance management, bioinformatics etc. The ORP-Sucking pests was an outcome of one such meeting, a project which contributed invaluable HRD into the national systems besides more than 60 international publications. Under his leadership the team developed DNA barcode for 50 different species of insect-pests of different orders; designed Species-Specific Markers for major insect-pests of horticultural crops; Dr. Kumar worked at Bioversity International (2016-20) with a responsibility to lead research in South and Central Asia and played a key role in organizing The first International Meeting on Agrobiodiversity (2016), two international meetings on Management of Tropical Race-4 of Banana (China 2018 and India 2020). Very recently (2020) as Vice-President of the Entomological Society of India, he organized two meetings on Biopesticides: Registration & Quality Assurance and Onion crop and Seed Production in relation to Onion Yellow Spot Virus. At Bioversity International he initiated research on development of an ABD index and valuation of Ecosystem Services in Collaboration with ICAR.

As a recognition of his scientific and managerial abilities Dr. Kumar, an entomologist was selected as DDG (Horticultural Science), ICAR in 2012. The PG school for Horticulture at IIHR, Bengaluru, Directorate of Floriculture at Pune besides innumerable scientific meeting etc. speak for themselves. Under his tenure Horticulture surpassed Grain production is an understatement.

He is associated with many national and international professional societies in various capacities. He is the Vice-President of Entomological Society of India. He is a recipient of several awards and honors. Primarily, he is the honorary fellow of Karnataka State Science and Technology (KSTA), recipient of Nammalvar Award for Conservation of Biodiversity, Dr Vasanthraj David Award for work in Entomology, Dr Roy Choudhary Memorial award from Phytopathological Society of India, Fellow of Society for Promotion of Horticulture, IIHR, Bengaluru, and many more. He is the research advisory committee (RAC) member of Central Silk Board, Chairman, RAC Indian Institute of Spice Research (IISR), Calicut, Central Tuber Crop Research Institute (CTCRI), Tiruvananthapuram, Directorate of Cashew Research, Puttur, Sericulture Biotechnology Research Laboratory, Kodati, Bangalore etc.

Dr Kumar has several national and international (110) research papers to his credit. Besides, he has written 5 books/book chapters. Dr. Kumar has edited a book on "Advances in IPM for Horticultural Crops", and an ICAR publication The Onion.

Dr. Surinder K Tikoo
Co-Founder and Research Advisor
Tierra Agrotech Ltd, Pune

Moderator

Dr Surinder K Tikoo, currently Co-Founder and Research Advisor, Tierra Agrotech Ltd, Pune, has five decades of experience as a successful vegetable breeder, teacher, and research & development manager. He has held global leadership positions over long periods in two private sector organizations, as well as spent 20 years with the Indian Council of Agricultural Research (ICAR) as a vegetables breeder, teacher, and research guide. He has developed several nationally and internationally released varieties and hybrids of tomato and okra, some of which are market leaders in their respective segments. He has served on several advisory committees of the Government of India and various professional bodies. Dr Tikoo, a PhD from Bangalore University, has published over 50 research and review papers. He also holds one patent. He has been conferred the Life Time Achievement Award for Excellence in Translational Research, jointly by the Society of Plant Biochemistry & Biotechnology and Springer.

Mechanisms of Resistance to Thrips and Thrips-transmitted Tospo Viruses and the Role of Crop Wild Relatives (CWR)

WebCon Series 21

N.K. Krishna Kumar
Former Deputy Director General (Hort. Sci.)
ICAR, New Delhi,
Former Regional Representative (Bioversity International)
South & Central Asia
Former Director, NBAIR, Bangalore

ATPBR
Breeding for Prosperity

ADVANCED TRAINING IN PLANT BREEDING (ATPBR)
A15, Amrut Sai Sara City, Nakshtrawadi, Paithan Road, Aurangabad, Maharashtra-431001

nipa®

NIPA® GENX ELECTRONIC RESOURCES & SOLUTIONS P. LTD.
New Delhi-110 034

ADVANCED TRAINING IN PLANT BREEDING (ATPBR)

NIPA® GENX ELECTRONIC
RESOURCES & SOLUTIONS P. LTD.

101,103, Vikas Surya Plaza, CU Block
L.S.C.Market, Pitam Pura, New Delhi-110 034
Ph : +91 11 27341616, 27341717, 27341718
E-mail: newindiapublishingagency@gmail.com
Website: www. nipabooks.com

For customer assistance, please contact
Phone: + 91-11-27 34 17 17
Fax: + 91-11-27 34 16 16
E-Mail: feedbacks@nipabooks.com

ISBN: 978-93-58876-50-5

Published and Exclusively Distributed by NIPA

"This WebCon is compiled and transcribed by Dr. M.J. Vasudeva Rao, currently a member of the Training Advisory Board of ATPBR"

Edited by Dr. Aruna T. Kumar

Composed and Designed by NIPA®.

Preface

ATPBR is bringing eminent plant breeders who have achieved laurels in not just science of plant breeding, but have excelled in other related areas like policy, administration, etc., to meet practicing and potential plant breeders on a platform called ATPBR Webinar series regularly every month. These web-based conferences have been hugely successful as seen by the numbers of participants from India and abroad. The WebCon series book captures the presentations in a published format which includes key messages delivered by the speakers.

Contents

Mechanisms of Resistance to Thrips & Thrips-transmitted Tospo Viruses and the Role of Crop Wild Relatives (CWR)

Dr. N.K. Krishna Kumar

Thank you, Dr Tikoo, for a nice introduction and thanks to ATPBR for inviting me to deliver this WebCon. I would like to start with a small introduction that this particular talk is structured as a general introduction to thrips and thrips-transmitted viruses. There is a bit of interest in everyone with regard to the black thrips which has caused havoc in the last couple of years. The topic has significant importance to plant breeders with regard to the new invasive Western Flower thrips, one of the most dreaded thrips, across the globe and the problems associated with the management of resistance in protected cultivation, and, finally how do we go about with regard to the entire management of the thrips and thrips-transmitted virus problems.

My talk today is basically about the mechanisms of resistance to thrips and the thrips-transmitted viruses and the role of crop wild relatives.

What is thrips?

- ✓ Minute insects, usually 1-2 mm in length.
- ✓ Some species up to 14 mm.
- ✓ Wings, each with a long fringe of marginal cilia, from which the name Thysanoptera is derived.
- ✓ Thrips are left handed i.e. only the left mandible is present--the most remarkable feature of thrips.

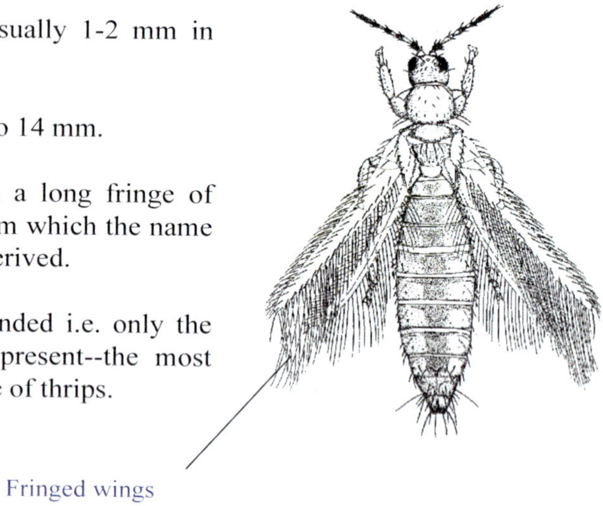

I would like to very briefly talk about thrips for those of you who may be from different sciences. Thrips are very small tiny insects, one to two millimetres in size and are a devastating pest across a range of crops especially vegetable crops, floriculture crops, rice and many other crops. The important characteristic feature of thrips, which I have shown in the diagram, is that they have fringed wings of marginal cilia from which the name Thysanoptera or thrips is derived.

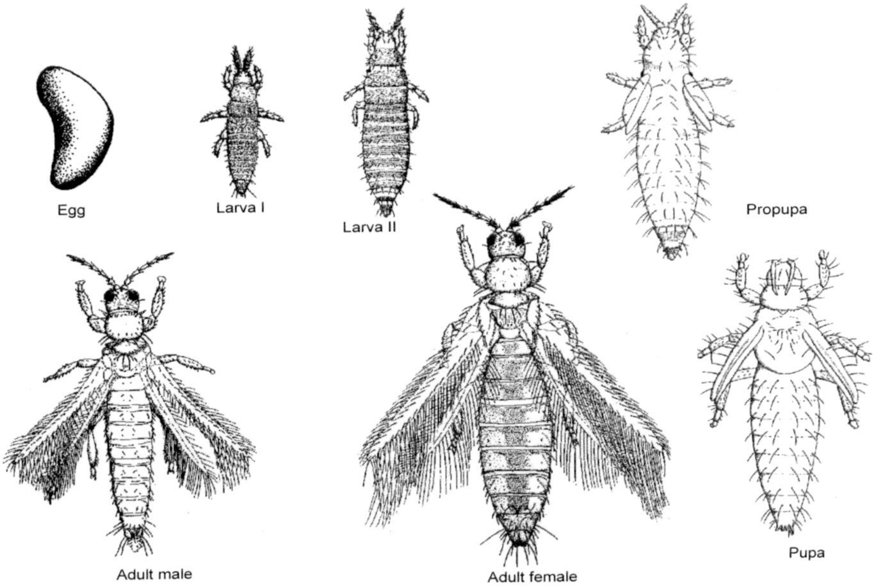

Before we go further, let me touch upon the fact that there is no true pupal stage in thrips. There is an egg which is often laid inside the epidermal tissue of the plant. Sometimes it is laid on the ovary. Larval stages almost mimic the adult thrips but for the absence of wings. There are two pseudo-pupal stages. We call it pseudo pupa because in the normal pupal state it does not resemble the adult, but in the pseudo-pupal state, the pupae also partly resemble the adult except that they are immobile and mostly found in the soil. The males and females are distinct, and usually the female is bigger than the male and you can see the ovipositor.

When we talk about thrips, the thing that comes to mind for most of us are the chilli thrips. I have shown here two plants, side by side - the one in my right hand is the one where the classical leaf curl with upward leaf rolling is seen which is caused by the thrips, whereas in the genotype in my left hand, the leaves are very normal. The thrips that you see on the right side of the slide is *Scirtothrips dorsalis* which is called as the chilli thrips.

Damage to chilli is a direct damage depending on which season and variety we are growing. The loss in chilli can easily be from 20 percent to almost the entire crop being burnt out if adequate control measures are not taken up at the right time. So, what I am really trying to focus through this slide is that there are certain crops where the chilli thrips can cause significant yield loss but it is not recognized many a times because it is also sometimes associated with tea mite or tarsonomid mite damage.

Thrips alone can cause very significant loss in chilli but due to the associated occurrence of mites or sometimes viruses or other insects, it is very difficult for a farmer often to clearly identify where exactly is the loss coming from.

The centre of origin of *Capsicum annuum*, commonly known as sweet pepper or chilli pepper, is Latin America. However, it has got two divergent evolutions since Columbus introduced it to other parts of the world. The primary centre of domestication was in Europe where the sweet peppers developed and, two, the Indian subcontinent or South Asian subcontinent, considered as the secondary centre of domestication where hot peppers or chilli peppers developed. Though there are two secondary centres of domestication from the same species, what is significant is that in case of bell peppers or sweet peppers, the loss due to thrips is significantly high, quantitatively and more so qualitatively, because there is a lot of cosmetic value loss associated with the marketing of sweet pepper.

The point that I am again indirectly trying to focus here, which I will come back at the end of this presentation again is that when we are looking for resistance in a particular crop or a particular species, what traits should we be really looking for? In case of chilli, we may have to focus more on quantitative loss, but in case of bell pepper, we have to extensively focus on the qualitative loss in addition to quantitative loss.

You may see the number of thrips that can practically appear on the surface of a sweet pepper fruit when weather is conducive.

But, when this kind of thrips damage occurs at young stage on a bell pepper crop, we can be absolutely sure that total yield loss can occur. Bell peppers are also at least two to three times more susceptible compared to chillis because of their secondary evolution which has taken place in Europe where they do not have the kind of thrips pressure that we notice in the tropics.

These are some of the peppers that were grown in net house. Bell peppers now are cultivated either in open field and under protected cultivation. If we can keep thrips away in protected cultivation, it can minimize significant yield losses, both quantitative and quantitative. But if it is a defective polyhouse, there will be tremendous problems associated with qualitative yield loss due to thrips. This slide shows a polyhouse which is reasonably well maintained.

When we are breeding for resistance with the red- and yellow-coloured *Capsicums*, especially from different seed companies, which are extensively cultivated mostly under net house, we may have to look at those parameters of thrips resistance which help us to cultivate bell pepper varieties which are free of thrips damage, in particular, from a quality perspective because they have tremendous cosmetic value loss. Long back, we did some experiments at IIHR on developing thrips damage rating scales to help chilli pepper breeders. We are all aware how difficult it is to accurately count thrips numbers on a leaf. By the time we complete counting on certain leaf area, we miss many of them. We, therefore, developed a thrips damage rating system at IIHR, wherein we correlated the thrips numbers collected in glass vials with thrips damage levels, and developed the rating system. Zero rating is the crop which apparently looks healthy. One rating is where the terminal leaves are showing a little bit of crinkling and in rating two the small bottom leaves are also showing mild crinkling. In three rating, there is upward leaf curling of almost all the top leaves. The four and five ratings are more severe damage levels as you can see in the slide. The ideal stage to contain thrips damage in a crop is when the crop is between one to two rating stages. At that stage, in case of chilli peppers, if sufficient interventions are taken up, significant yield loss can be prevented. One of the easiest methods of controlling thrips is by taking control measures at the early stage.

In addition to directly damaging the host crop by sucking the leaf sap, thrips also damage by being vectors of disease-causing viruses, for e.g., groundnut bud necrosis virus (GBNV). The three most classical symptoms of insect-transmitted viruses are – firstly, the effect of the virus is especially seen in the apical region. The bottom leaves will be green and the terminal portions of the plant will be wilting and in a very short time the entire plant will look burnt out. Secondly, the insect-transmitted viruses are sporadically distributed in the field at the initial stage, i.e., one plant is showing virus symptoms and the adjacent plant may be looking perfectly healthy. Thirdly, by any amount of insecticide drenching or any other similar measures, the plant will not be able to recover once it is infected by a virus. With the help of the modern molecular techniques, we can identify exactly the type of virus causing the disease.

One of the most important problems today which the vegetable industry is facing is the incidence of the black thrips. This is *Thrips parvispinus*, a name which I myself had not heard about three years back. This was reported for the first time in India by my SRF, Dr Vikas Kumar and Dr. Kumud Tyagi, who reported it on Papaya at Bangalore for the first time. They also cautioned in 2012 that this is a thrips which needs to be closely monitored. In spite of that caution, probably, in my opinion, and, this is also for the seed industry to note, that in our country the quarantine is not what it should be. Indiscriminate import of flowering plants and other planting material without scientific quarantine needs to be avoided. This insect is suspected to have come from Dendrobium or other plants which are imported from Thailand and Indonesia probably somewhere close to Kadiyam and from where they spread to Guntur and other regions. Today we have a serious problem on our hands. Something needs to be done to the quarantine in our current times. Once the problem escalates, IPM alone to manage this problem may not be possible from a plant protection perspective.

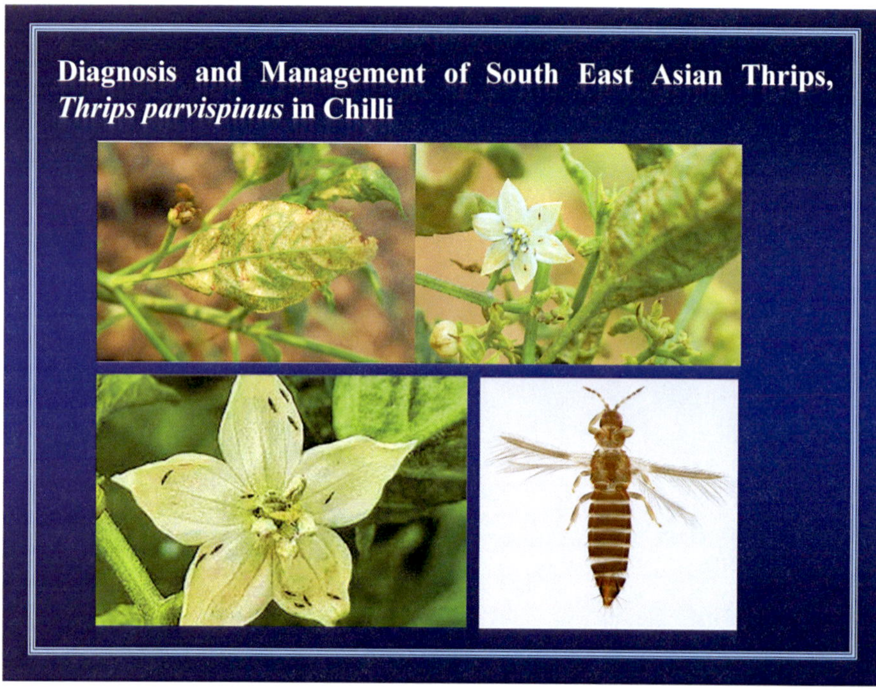

This thrips is called black thrips and has two characteristic features. Firstly, unlike the other thrips, this thrips congregates on the floral petals in significant numbers, from 10 to 75. Secondly, the lower surface of the leaves is concentrated with nymphs, whereas the adults feed both the upper and the lower surface of the leaves. What this thrips is telling us is that there are two things - one is that these thrips are noticed immediately after heavy rain, and, second, by their appearance on the petals and on fully opened leaves, it appears to me, that rainfall does not have a significant influence on the multiplication of this pest. As a result, this pest is occupying a small niche area where *Scirtothrips darsalis* does not prefer to be there post rains. This is something that needs to be researched to find out before we make serious attempts at its management. *Thrips parvispinus* as you can see is typically blackish and very much present on the petals. So, if there are thrips hovering around the petals and they are black in colour, it is probably, *Thrips parvispinus*.

In India, as per reports in 2021-22, nine lakh acres of chilli crop in South India was affected. Surprisingly, this was never anticipated. Sometime back when I was talking with the private sector on plant protection, they said this pest is becoming a real problem in Northern Karnataka and parts of Andhra Pradesh also and not satisfactorily controlled through insecticides.

When this problem was noticed, Government of India issued a letter indicating what needs to be done. However, what is needed on the ground is that the nation should stand in preparedness of such problems, and anticipatory research investment is needed with a dedicated research team which is working round the clock. I don't think these problems can be handled through directives or selecting couple of centres to come out with the magic answers for such problems.

Dominance of *Black thrips over* chilli thrips of India: A likely case of species displacement

Thrips Species	Andhra Pradesh	Karnataka	Telangana
T. parvispinus	96-98	94-96	96-98
T. florum	1	1	1
T. hawaiiensis	2-3	3	2-3
T. palmi	<0.5	<1	0
Scirtothrips dorsalis	0	0	0

Mechanisms of Competition Leading to Displacement. Most displacements identified were triggered by the introduction or invasion of an exotic species (78%), although other environmental factors may predispose a species to being displaced. Polyphagy and the lack of natural enemies often characterize invasive species that have large impacts on communities. Such invasive species typically have their greatestimpact on simple and anthropogenically disturbed habitats (97).

Sridhar *et al.*, 2021

We can see in the above table that in matter of a short time, *Thrips parvispinus* percentage was anywhere from 94 to 98 percent in Andhra Pradesh, Karnataka and Telangana, whereas the normally associated *Scirtothrips dorsalis* was absent during the time of survey. The black thrips is seen displacing *Scirtothrips dorsalis* in the post-October cropping period, i.e., during November, December and January, and causing significant crop damage. The reasons for this could be many, but at this point of time there are many unknowns. We have also not fully understood, how this species of thrips has been able to damage the chilli crop significantly. It is also present on sweet pepper, egg plant and a range of other crop plants.

Table 2: Host plants ot7.parvispinus recorded from Andhra Pradesh

Name of the host plant	**Family**
Chi 11 j Capsicumannuum L	Solanaceae
Sweet pepper, Capsicumannuum L.	Solanaceae
Brinjal, Solanum melongena L	Solanaceae
Black gram, Vignamungo L.	Fabaceae
Pigeon pea, Cajanuscajan(L) Millsp	Fabaceae
Common bean, Phaseolusvulgaris L.	Fabaceae
Bitter gourd, Momordica charantia L.	Cucurbitaceae
Ivy gourd, Coccinia grandis (L.) Voigt	Cucurbitaceae
Watermelon, Citrul 1 us lanatus (Thunb.) Matsum. & Nakai	Cucurbitaceae
Name of the host plant	**Family**
Cucumber, Cucumis sativus L.	Cucurbitaceae
Bottle gourd, Lagenaria siceraria(Molina) Standi	Cucurbitaceae
Drumstick, Moringa oleiferaLam	Moringaceae
Coriander, Coriandersativum L	Apiaceae
Marigold, Tagetessp.	Asteraceae
Chrysanthemum, Chrysanthemum indicum L	Asteraceae
Jasmine, Jasminum sp.	Oleaceae
Mango, Mangifera indica L	Anacardiaceae
Cotton, Gossypium sp.	Malvaceae

Evaluated entomopathogens- fungi, epns,
Evaluated insecticides
Evaluated plantoils and is in progress

The black thrips has also been reported on other crops. In Cotton also it has been reported. I will not go into the list which is given in the above two slides. But what is astonishing is that the proportion of females to male ratio in certain crops shows that there is a preference for female thrips as compared to the male thrips in crops such as cotton and others.

Distribution of *T. parvispinus*

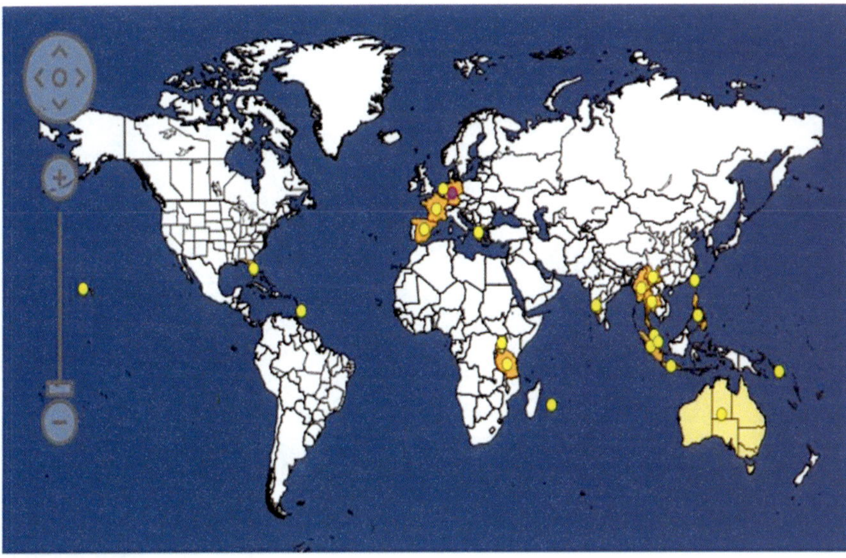

Lengend: ○Present ●Transient

The black thrips is not just unique to India, it was very much there in Southeast Asian countries. It was not only causing extensive losses in Indonesia and Malaysia, but also in Hawaii. What we are inferring as invasive species again and again is that most of the invasive species are originating through trans-pacific pathway than the trans-Atlantic. We need to really strengthen our quarantine systems if we have to prevent the entry of many of these pests. This pest is also reported in China and the illegal trade, illegal importation and lack of stricter quarantine and inability to report them on time is making these quarantine pests not to be noticed before they flare up. We need to learn lessons.

Southeast Asia

- *Thrips parvispinus*, a member of "Thrips orientalis group" (Mound 2005), is a widespread pest thrips species of quarantine importance
- Identification of an invasive pest is the first step which unlocks the barriers for further research in planning appropriate management strategies for the pest involved.

- **Distribution data**
  - **Known from Thailand and Malaya to New Guinea and northern Australia, also Hawaii, Micronesian Islands, and Greece … India…….**
  - ***Thrips parvispinus* (Karny)**
  - **Original name and synonyms**
  - ***Isoneurothrips parvispinus* Karny, 1922**
    ***Isoneurothrips jenseni* Karny, 1925**
    ***Isoneurothrips pallipes* Moulton, 1928**
    ***Thrips (Isoneurothrips) taiwanus* Takahashi, 1936**

Thrips parvispinus was reported by Lawrence Mound who is from Australia. He is also the world authority on thrips. For those of you who want to understand thrips better, please buy the books on thrips by Lawrence Mound. I think it will do a world of good to all the seed companies if you can have this book on hand.

Host Range of Thrips parvispinus –

Causes Direct feeding damage.

Most affected: Papaya, peppers, potatoes, eggplants, beans, shallots .

In Indonesia, field pepper 23-65 % loss (Johari et al. 2014).

Ornamentals : *Dahlia, Chrysanthemum, Gardenia, Dipladenia, Ficus, Anthurium, Hoya etc*

In Malaysia, on papaya is & Secondary infection by saprophytic *Cladosporium oxysporum* (bunchy and malformed top) -

Extensive leaf damage on Gardenia plants in Greece.

Capsicum in Java,

Vegetable crops in Thailand.

Vector of tobacco streak ilarvirus in transmission studies from infected tomato pollen to seedlings of *Chenopodium amaranticolor.*

Hosts varying across its geographic distribution.

Papaya is one crop which the black thrips prefer and is most affected by it. In Indonesia it is on peppers, ornamental crops and Garcenia. I am not sure how far it is true that there is one stray report of it as a vector of tobacco streak Ilar virus. Ilar viruses are a group of viruses like sunflower necrosis virus. When thrips feeds on the contaminated pollen and rubs it onto other leaf surfaces, it is able to infect the plant. Most of you may know to what extent we have taken a hit from the sunflower necrosis, which is an Ilar virus. Sunflower production came down significantly because of thrips-facilitated transmission of this virus.

Reports of *Thrips parvispinus* in Indian States and UTs as on 28/1/2023

Review Article
The J. Res. PJTSAU 49(4) 1-11, 2021

GLOBAL STATUS OF *THRIPS PARVISPINUS* (KARNY, 1922), AN INVASIVE PEST

V. SRIDHAR[1], P. SREE CHANDANA[2] and R.R. RACHANA[3]
[1,2] Division of Crop Protection, ICAR-Indian Institute of Horticultural Research, Bengaluru - 560 089
[3] ICAR-National Bureau of Agricultural Insect Resources, Bengaluru - 560 089

Date of Receipt : 11-12-2021 Date of Acceptance : 20-12-2021

ABSTRACT

Thrips are important group of sucking pests which cause significant economic losses both as pests and vectors of serious plant viruses in several horticultural crops. Recently an outbreak of *Thrips parvispinus* has been reported from southern states of India (Andhra Pradesh, Karnataka and Telangana) especially on chilli crop causing 70-100 per cent damage. *T. parvispinus*, a member of "*Thrips orientalis* group", is a widespread pest species of quarantine importance and designated as one of the pest species of South East Asia. *T. parvispinus* has displaced *T. palmi* in Indonesia indicating its competitive ability in displacement of other species in the crop ecosystem. Successful quarantine interceptions made throughout the world against this pest shows the importance of interceptions in avoiding the entry of invasive pests in to any country. Despite existence of the quarantine provisions, recent invasion of various exotic pests like South American tomato moth, rugose spiralling white fly, fall armyworm etc. into India in quick succession is a concern particularly under globalisation situation. Considering the seriousness of the damage caused by the invasive pest in India in the recent past, an attempt was made in this review to present the status of *T. parvispinus* at national and global level including its identification, taxonomic status, host range, development and biology, extent of damage and various management strategies. Importance of various IPM tools to be explored for the management of this invasive thrips is also discussed in the review.

Keywords: Chilli, invasive, *parvispinus*, polyphagous, thrips

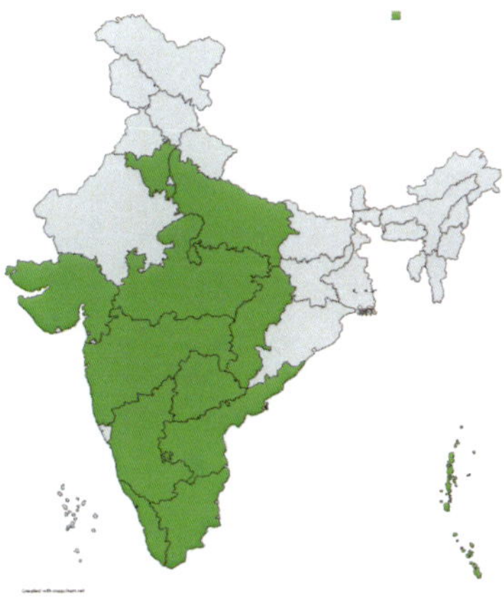

In all the states shown in green on chilli except Kerala where it is reported on mango.
T. parvispinus in 1 : 10-15 ratio of male and female

Thrips parvispinus was considered as basically a South Indian problem but it should not be taken for granted, as in just a matter of short time, it has potential to spread to many other parts of the country. The type of damage from black thrips is very difficult to see externally. It is difficult to differentiate from *Scirtothrips dorsalis* except that it is on the flowers as well as it appears during post monsoon period. Yield loss in many places was to the tune of 85 percent.

Nature and Symptoms of Damage

- ✓ *Thrips parvispinus* adults colonize mainly on flowers and underside of leaves while larvae restrict themselves to under surface of the leaves.
- ✓ Both adults and larvae damage plants by rasping and sucking of the plant sap. Heavy infestation affects the growth of the plant, flower drop and reduces fruit set and development, ultimately resulting in yield loss.
- ✓ The symptoms due to thrips damage include deep punctures and scratches on underside of the leaves. Infested undersurface of the leaf turns reddish brown, whereas upper side of such leaf looks yellowish.
- ✓ On floral parts, brownish streaks appear on petals due to scraping by thrips. The damage results in drying and withering of flowers leading to reduced fruit set.

As I said earlier, black thrips feed mainly on the flowers leading to flower drop as well as the infested under surface of the leaf turn reddish. Upper surface looks yellowish and on floral parts brown streaks also appear which does not appear in case of the chilli thrips.

Table 1. Spearman correlation coefficients and significance between damage score in screening methods of thrips resistance in pepper.

		T. parvispinus		*F. occidentalis*			
		Leaf disc	Detached leaf	Glasshouse	Leaf disc	Detached leaf	Cutting
T. parvispinus	Screen house	0.77 ***	0.80 ***	0.76 ***	0.65 **	0.70 ***	0.53 *
	Leaf disc		0.87 ***	0.71 ***	0.71 ***	0.71 ***	0.45 *
	Detached leaf			0.73 ***	0.70 ***	0.69 ***	0.50 *
F. occidentalis	Glasshouse				0.77 ***	0.73 **	0.48 *
	Leaf disc					0.77 ***	0.41
	Cutting						0.64 **

Top figure: correlation coefficient
Bottom figure: significance. *, **, and *** indicate significance at P<0.05, P<0.01, and P<0.001 respectively

Screening tests were carried out by taking small leaf discs of different genotypes and allowing black thrips to feed on them. There is a good amount of feeding which corresponds to the damage levels. Dynamic scoring was carried out to identify which genotype were preferred or which genotypes were not preferred. Correlation coefficients were obtained by simple leaf feeding experiments and damage scores. We may need mass culture facilities for black thrips at this point of time. But certainly, it requires many more experiments and we need a dedicated culture of *Thrips parvispinus*.

Management at global level

- Information is very meagre
- Spinosad is better than acetamaprid (Murai et al., 2010)
- In India *Alecto* and *Delegate* give 60% control
- White traps are better than blue traps
- A few resistant lines Maharijaya et al., 2011
- Metarhizium and verticillium

Information on the management of black thrips is very meagre. All kinds of chemicals are recommended and one cannot be sure about any specific chemical for control. We not only have an invasive pest in black thrips, but also have a thrips which is resistant to insecticides. How this is happening on a pest which has not been exposed to that amount of insecticide pressure is still baffling. A few resistant lines in chilli were reported by Maharijaya in Indonesia.

Integrated Pest Management (IPM) approaches for *T. parvispinus*

- ✓Good Nursery management under nylon nets
- ✓**Seed treatment with Imidacloprid 70WS @10 g per kg seed.**

- ✓ Border cropping in main field with Maize, Sorghum or Bajra (2-3 rows) 30-40 days before transplanting

- ✓ Clean cultivation by clearing alternate hosts and crop residue and maintain weed free bunds.

As we know there are other aspects which are being recommended for control of black thrips. As usual they include application of Imidacloprid seed treatment, border cropping, etc. With all these general kinds of recommendations, I don't think, at this point of time, we are able to achieve satisfactory control of black thrips.

- ✓ Use of *Metarhizium anisopliae* or *Pseudomonas fluorescens* @ 2 kg/t along with recommended doses of farm yard manure (25 to 30 t/ha).
- ✓ Soil application of 500 kg of Neem cake /1.50 - 2.00 ton of vermi-compost/ha.
- ✓ **Seedling root dip for 30 minutes with Imidacloprid 17.8% SL @ 0.5 ml/ L before transplanting.**

The one thing which I can say is that the black thrips are reasonably attracted to colour sticky traps. There are some farmers who say that it is possible to trap them but I have my own doubts. Because in nature there are billions of thrips and if someone says that I can trap thrips using this or that, it is something which I would not agree.

✓ Follow recommended spacing as per the cultivar. Avoid closer spacing.

✓ **Erection of white or blue sticky traps at 65–75 traps/ha.**

✓ Drip irrigation and Mulching with silver coloured polythene sheets of 25–30 micron thickness to reduce pupation of thrips in the soil.

✓ **Frequent inter cultivation (earthing up/raking of soil) operations to destroy soil inhabiting pupae of thrips**

On initial occurrence of thrips in field

- **Spraying 4 % Neem Seed Kernel Extract (NSKE) or 5% Neem Seed Powder Extract or 0.50% Pongamia oil (5 ml/L)**

But what I think is important is to know whether it is a vector or not of the disease-causing viruses.

When the incidence is high

- **Spinosad 45 % SC @ 0.25 ml/l**
- **Spirotetramat 15.31% OD @ 0.75ml/l**
- **Dimethoate 30 % EC @ 2ml/l + Lambda-cyhalothrin 5% EC@ 1ml/l**
- **Emamectin Benzoate 01.50 % + Fipronil 03.50 % SC @ 1.25ml/l**
- **Diafenthiuron 47 % + Bifenthrin 09.40 % w/w SC @ 1.25 ml/l**
- **Rotation of insecticides with different modes of action groups eg. Spinosad-imidacloprid- Fipronil**
- Drenching with chemicals like chlorpyrifos 20 EC @ 5 ml/L or spreading of neem cake on beds or **Soil application of EPNs, *Steinernema carpocapsae* or *Heterorhabditis indica* @ 5 kg/Acre.**
- **Pongamia oil and neem oil @ 2 ml/L along with insecticide sprays once a month as synergists**

To control black thrips, whether to use Spinosad, or Dimethoate or Emamectin, etc., I will not go into this.

Immediate requirements met.....

Advising the farmers through adhoc recommendations

Government of India did some amount of extension on the management aspects of black thrips. However, at this point of time, we are still groping in the dark regarding black thrips control.

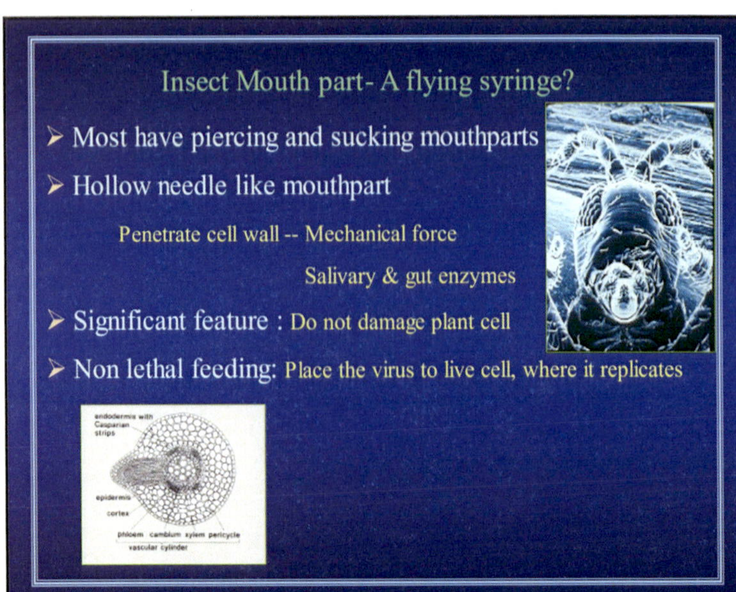

Having talked a little bit about the *Thrips parvispinus* damage, we will come at the end of this talk to the other aspects of management. Whatever be the damage - whether it is feeding damage or qualitative damage or damage to different crops or damage through its ability in transmitting virus, like any other sucking insects, thrips also have piercing and sucking type of mouth parts. Earlier entomologists said it has rasping type mouth parts. No! It has got piercing and sucking type mouth Parts. They can very well penetrate into the cell wall. They feed on the epidermis, but very rarely on the vascular bundle. Whenever they do transmit a virus, the damage does not happen significantly to the plant cell as a result the virus is able to replicate in the live cell and thereby cause the disease.

The iris yellow spotted virus on onion seed crop is another one. There are many virus diseases which are emerging, including capsicum chlorosis. When I did my PhD during 1989 to '93, tomato spotted wilt disease was thought to be the only tospo virus which was transmitted by thrips. The list now includes at least 20 to 25 viruses which are being transmitted by thrips.

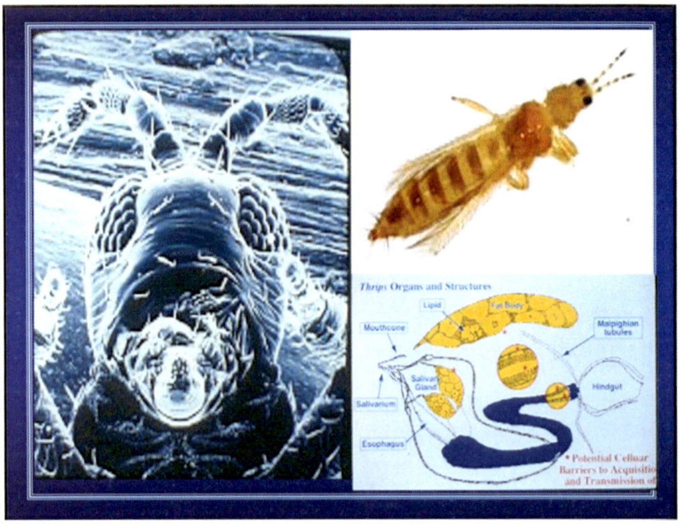

My advisor, Dr Ullman, did a lot of work with regard to the digestive system of the thrips. In case of thrips, we can see on the left side of the slide a close-up of the scanning electron micrograph of the thrip's mouth parts. On the right-hand side is the dissection and we can see that they have a yellow fat body and the salivary glands.

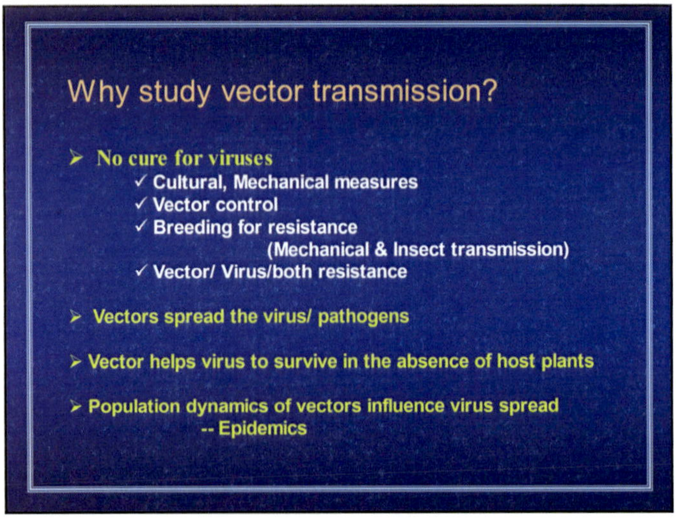

In case of thrips, we may ask if we have resistant planting material that should be enough, and why do we need to understand vector transmission. We all know that there is no cure for viruses, whether it is in plants or animals. We have very well realized this in the case of Covid-19 pandemic. No other global event shaped our destiny like Covid-19 did. There are two aspects that are very important whenever we are not able to manage a viral disease. Firstly, indiscriminate application of chemicals for control causes more harm than good, not only in terms of pesticide pollution, but also in terms of development of resistance. Secondly, it may create new biotypes which become more difficult to manage.

Often, we mechanically inoculate the infected sap and then screen for resistance. But, at the same time, in nature, nobody infects mechanically. It is only insect transmission that takes place. Let me give a simple example of tobacco mosaic virus which can be mechanically transmitted even by touching a contaminated leaf and no insect is able to transmit TMV.

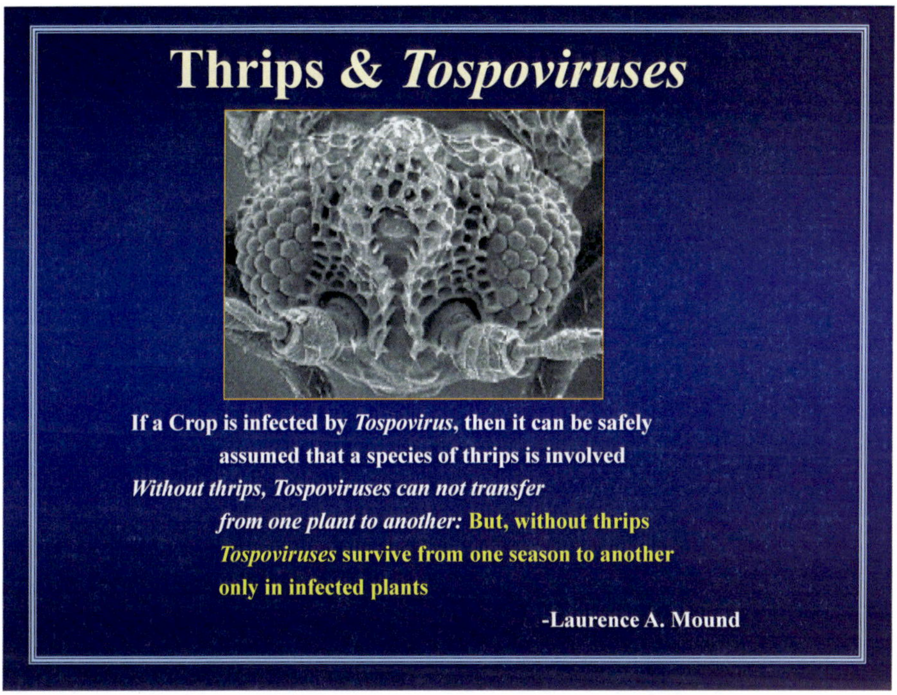

The evolutionary association between an insect and the virus is ancient in nature. No doubt we may have preliminary screening for resistances, but before we release a variety, it is absolutely necessary that insect transmission is studied as it happens in nature. And for which, many types of insect transmissions need to be understood. It is also necessary to have, for example, as in the case of TLCV, the *Lycopersicum hirsutum*, which is now called *L. hybrochitiss*, has conferred a great deal of resistance to insects, and TLCV. Therefore, if we can have a variety/hybrid which has resistance to both the vector and the virus, it will be excellent. While vectors spread the virus from one plant to another, the virus also survives in the absence of the host plants. The population dynamics of the vector determines the epidemics. I will quote Lord Laurence Mound who said that "if a crop is infected by a tospo virus, then it can be easily assumed that a species of thrips is involved. Without thrips, tospo viruses cannot be transferred from one plant to another. But without thrips, tospo viruses survive from one season to another in the infected thrips". So, these are some of the things as to why it becomes important to understand viruses, vector and the environment.

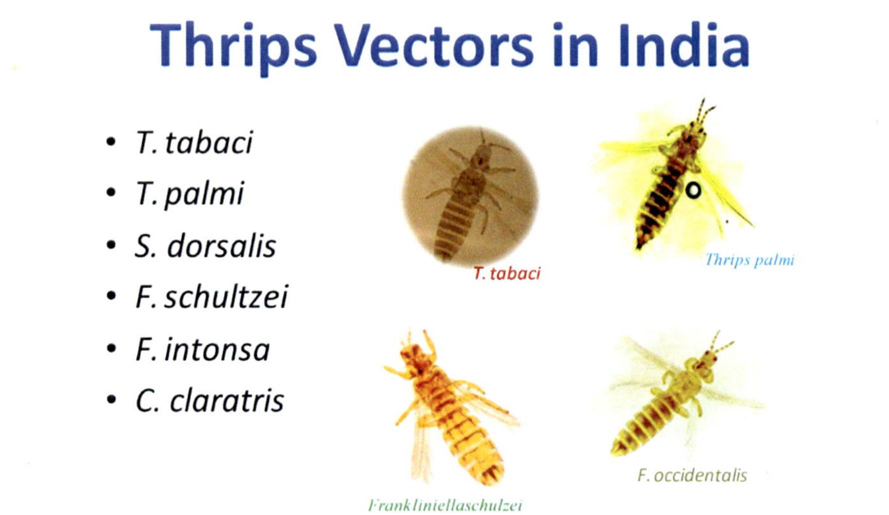

There are many species of thrips vectors in India and to that we have now added the *Frankliniella occidentalis*.

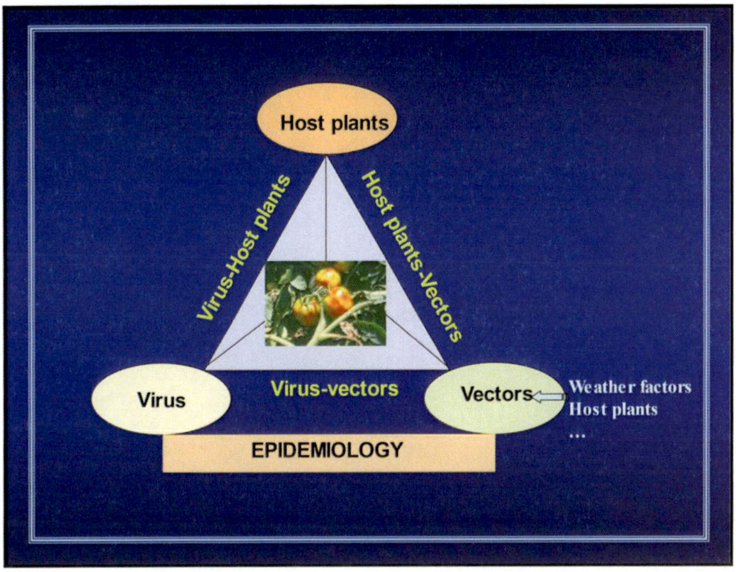

As it is, the management of tospo virus or any pathogen is difficult because of the interactions between the pathogen, the host plant and the environment. However, when a vector is also involved, there is a fourth dimension which intervenes where the vector itself is influenced by the weather, virus-vector interactions, environment, and of course, the host plant. So, we have a complex scenario which makes studying epidemiology very difficult.

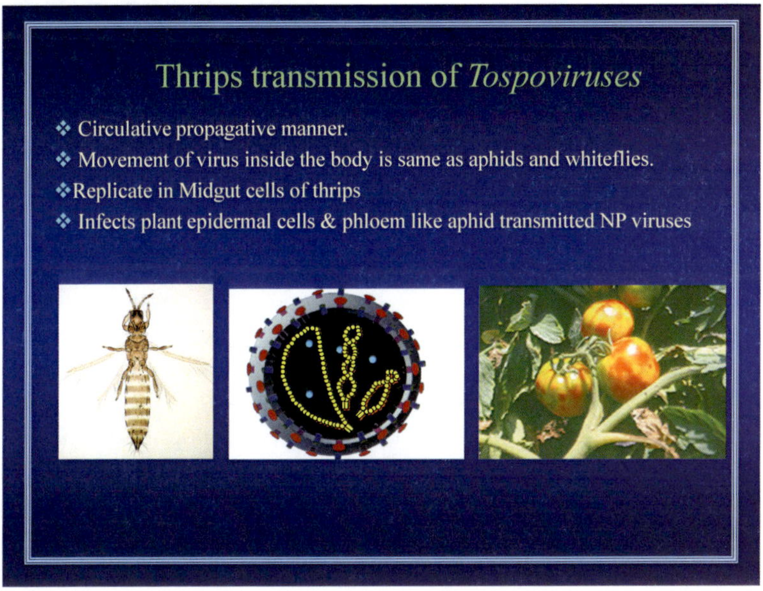

For thrips transmission of tospo viruses, the tospo virus move through the gut, exit the midgut, enter the salivary gland and multiply in the salivary gland of the thrips. This has got tremendous implications not only for the management of the disease, but also for quarantine aspects of the pest. If we import planting material and if there is a vector + thrips, then we have the problem of the infected vector spreading a new tospo virus into different set of plants in new areas.

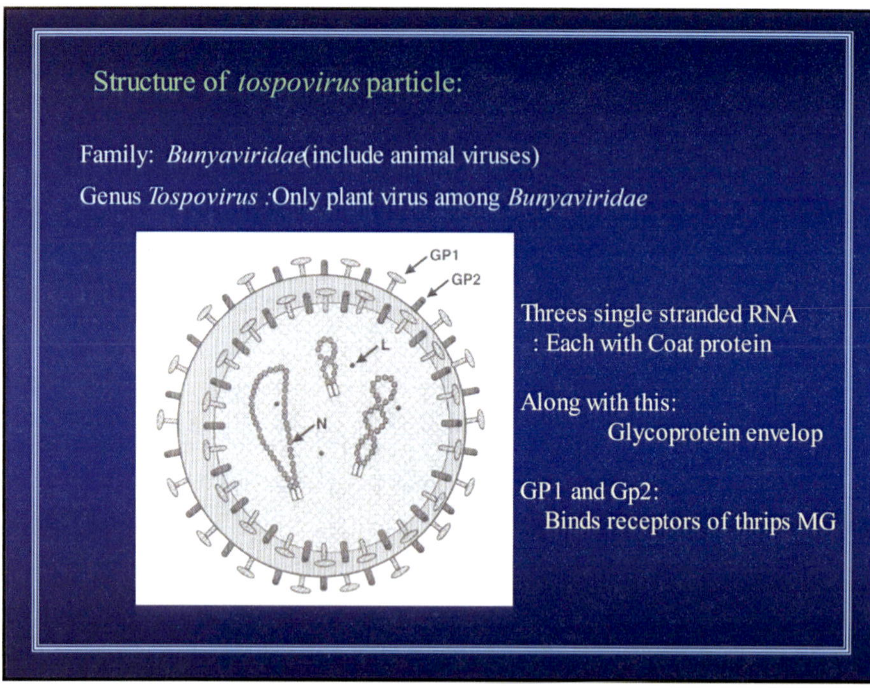

Tospo viruses looks somewhat like this. I think many of you have seen covid virus and its structure. This particular family of viruses are called *Bunyaviridae*. Earlier it was called as *Arbo* virus, for arthropod-borne viruses. All its cousin viruses are animal viruses. Tospo virus is the only plant virus from a group of viruses where other members of the group are all animal viruses like encephalitis, blue tongue and many others. It is a tripartite virus with a small, medium and large RNA. Like in the case of covid it has also got 2 glycoproteins spikes, but tospo is vector transmitted while the covid virus is not vector transmitted. In the case of tospo, the glycoproteins are the ones which are associated with their binding to thrips and transmission.

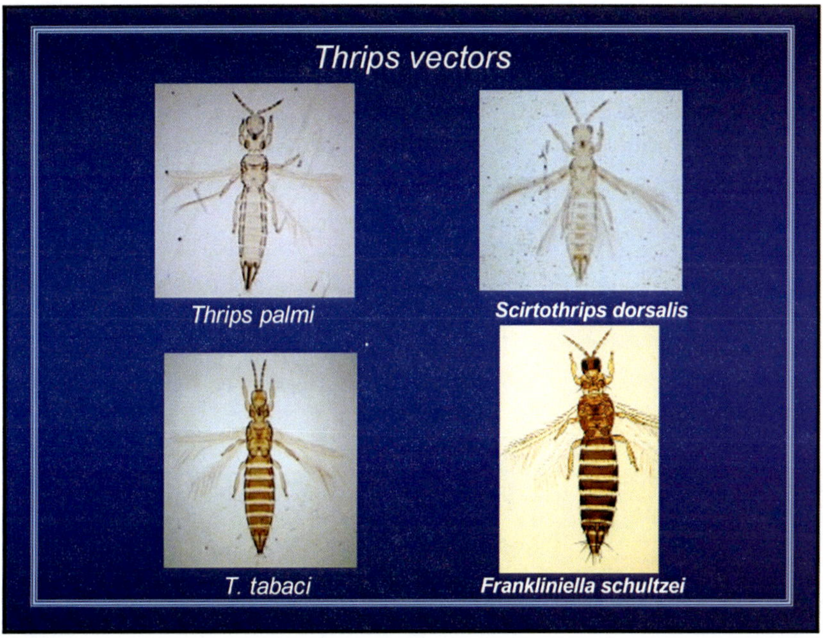

These are photographs of some of the thrips' species, which is self-explanatory and hence, I will not just go into this.

Sl. No.	Thrips	Tospo transmitted	Name
1	*Frankliniella occidentalis*	5	CSNV, GRSV, INSV, TCSV, TSWV, Melon severe Mosaic Virus (MeSMV)?
2	*F. schultzei*	5	CSNV, GRSV, GBNV, TCSV, TSWV
3	*F. fusca*	2	TSWV, INSV
4	*F. intonsa*	4	GRSV, INSV, TCSV, TSWV
5	*F. bispinosa*	1	TSWV
6	*F. zucchini*	1	ZLCV
7	*F. gemina*	2	TSWV, GRSV
8	*F. cephalica*	1	TSWV
9	*Thrips tabaci*	3	IYSV, TSWV, TYFRV
10	*T. palmi*	4	CLCSV, GBNV, MYSV, WSMV
11	*T. setosus*	1	TSWV
12	*Scirtothrips dorsalis*	3	GBNV, PCFSV, PYSV
13	*Ceratothripoides claratris*	1	CCV
14	*Dictyothrips betae*	1	Polygonum RSV
15	*Sericothrips variabilis*	1	Soybean Vein Necrosis Associated Virus (SVNAV)
16	?	1	Tomato Yellow Ring Virus (TYRV)

We can see that the tospo viruses transmitted by *Frankliniella occdentalis* is more than others. I would say that even *Thrips palmi* is no less. Unfortunately, in India, it is like Anopheles the vector of malaria and Aedes, the vector of dengue coming together. Now we have both *Frankliniella occidentalis* and *Thrips palmi* coming together and these two dreaded insect vectors are extremely polyphagous, extremely efficient, and transmitting a group of tospo viruses. We may have a problem on our hand, in future, if we don't address these at a basic level, i.e., at a molecular level, and also at an applied level. This is just to warn that we had a problem only with *Thrips palmi*, and now with *Frankliniella occidentalis* coming into the scene, the potential issues need more serious attention. In Europe and United States, the very name *Frankliniella occidentalis* or Western Flower thrips, is a matter of concern because of the number of viruses that they transmit

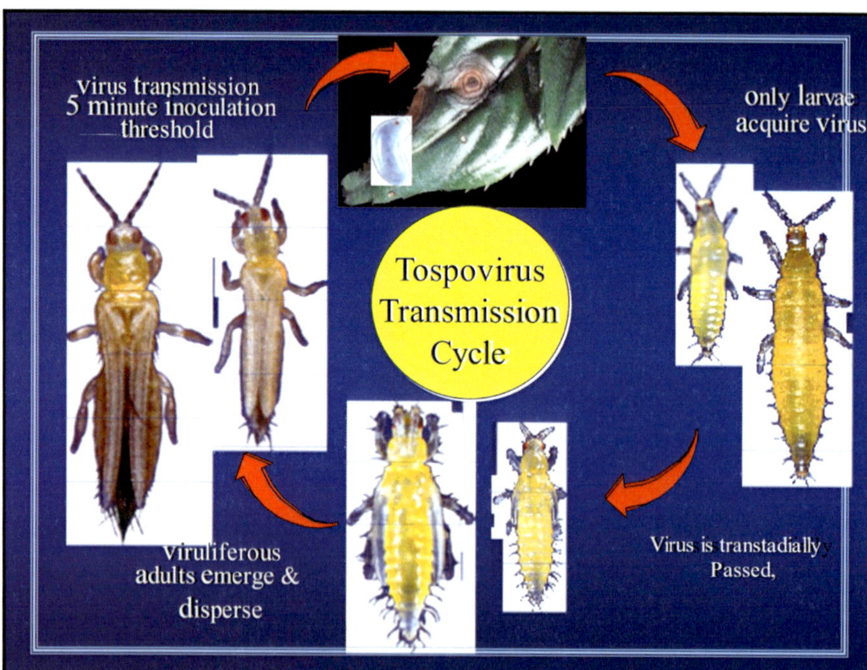

In case of thrips, another important feature we should know is that the tospo viruses are acquired up only by the larvae. That means only on an infected plant thrips will be able to acquire the tospo virus. Once they acquire the virus, normally larvae are not in a position to transmit the same, for two reasons – that they are very poor transmitters, and that they can't fly.

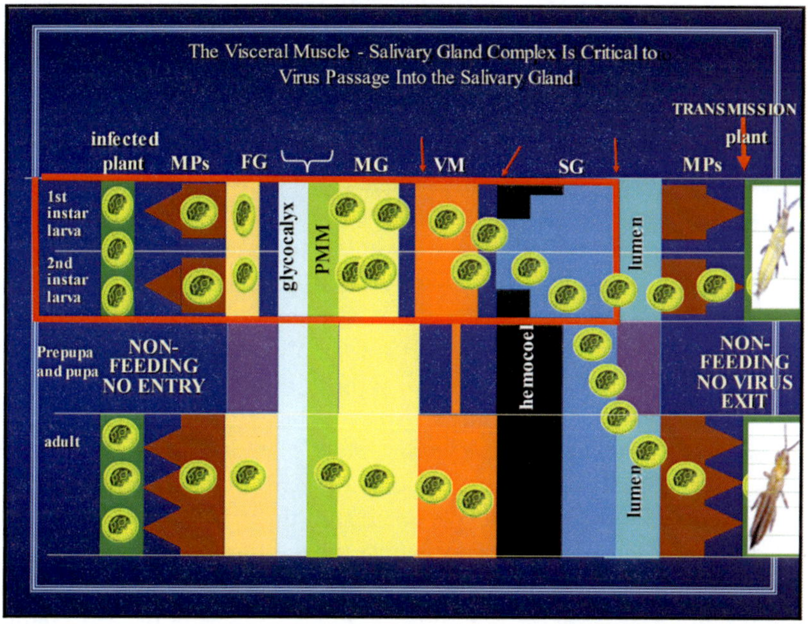

After the pseudo-pupal stage, the adults emerge, during dispersal, 10 to 15 minutes of probing is enough for it to transmit the tospo virus on to any suitable host plant.

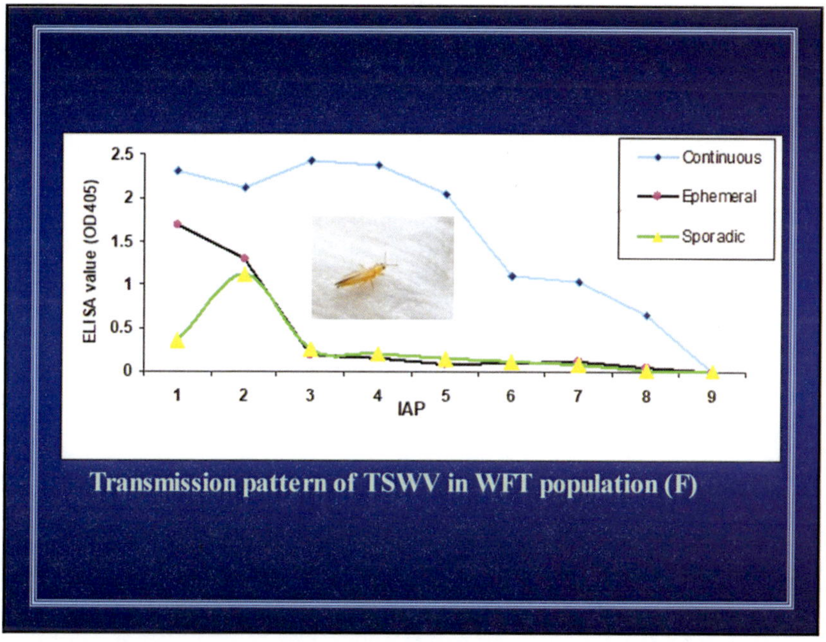

Now, these studies are important because we have host plants for thrips, but not susceptible to the virus. We have host plants which can have the virus but do not support thrips multiplication in large numbers. We may have host plants which are for both support thrips and tospo virus. We may have host plants which do not support thrips and tospo virus. So, in nature, it is a combination of all the above which drive the thrips transmitted transfer of viruses. I worked on some of these aspects during my postdoctoral work at UC, Davis. We often think that when we do some laboratory studies, we take a couple of thrips and we do the transmission studies. Some of these studies report results as positive transmission. Some say it is not a vector. Some say I did not get transmission. The transmission of thrips by tospo viruses is an extremely complex process. For example, in this case, 50 percent of the thrips are only transmitters and rest 50 percent did not transmit the virus. This was based on innumerable single-thrips studies. We carried out these studies in UC, Davis and shipped the thrips all the way to Wisconsin where they did the RTPCR to determine the rate of replication of the tospo virus. We correlated them with their transmission studies in UC, Davis. We observed that in those individual thrips where the TSWV replicate in large numbers, they transmit the disease almost all through their life as you can see in the blue line. In about 25 percent of the population where the TSWV replicate but not in great numbers, they transmit after adult emergence maybe for about three to four days after which they lose their ability to transmit. In some of them, they may, at one spike, transmit and after that they may lose their ability to transmit. What is important is the proportion of thrips that is there in nature which transmits all through their life, and what is the proportion that is ephemeral - is it in *Thrips palmi*, as it in Western Flower Thrips, or is it in other species? The fluctuations that happen ultimately determines the epidemics in nature.

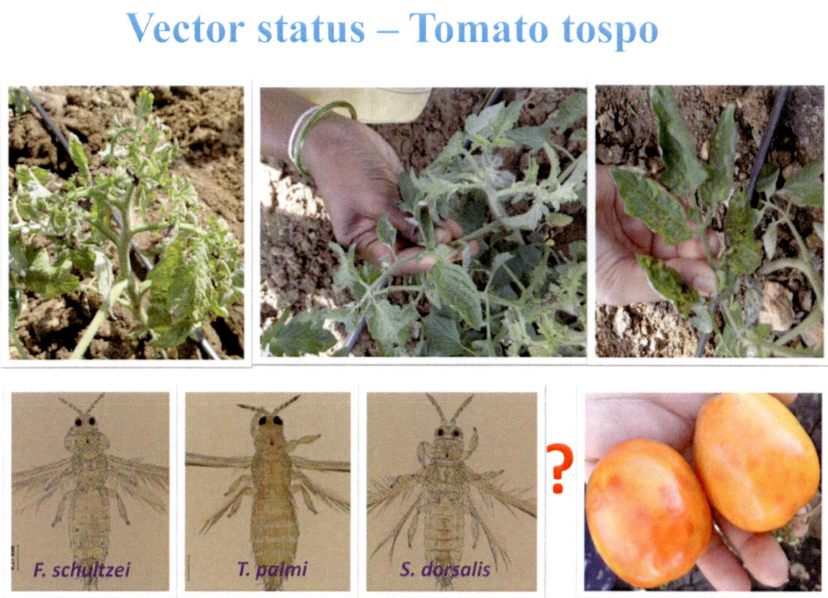

This is with regard to some of the damages that we may see with regard to the Tomato tospo.

This is the classical *Thrips palmi* which is a yellowish thrip. *Thrips palmi* is also a major problem both as a pest and as a vector in transmission of watermelon bud necrosis virus. *Thrips palmi* are typically called as melon thrips.

For those who are working on the watermelon bud necrosis management, the first thing to get is a culture of *Thrips palmi*. Then probably also have Western Flower Thrips for transmission studies. If we are working on chillies, we should look into *Scirtothrips dorsalis* and then Western Flower thrips. Western Flower Thrips either way come into the picture.

About eight to ten years ago, contaminated consignments of chrysanthemum were imported indiscriminately with the TSWV and Western Flower Thrips. The importers grew them in the polyhouse and did not realize that it carried the Western Flower Thrips transmitted TSWV. The chrysanthemum flowers were sold in market and both Western Flower Thrips as well as the TSWV began to spread and now it is there in Karnataka, Tamil Nadu and possibly in other states. I would urge entomologists working in different parts of the country and other parts of the world to closely look into the presence or absence of the Western Flower Thrips. Dr Asokan (IIHR) and Dr Vikas Kumar in Zoological Survey of India can do detailed molecular analysis and confirm whether the thrips that is observed is Western Flower Thrips or *Thrips palmi*.

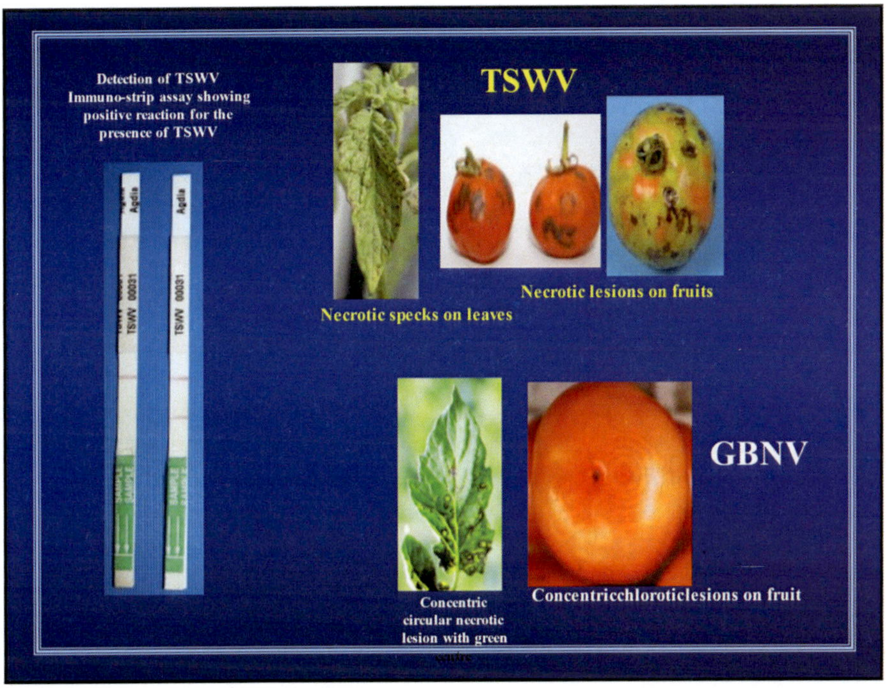

We now have a problem of TSWV and Western Flower Thrips. I also urge upon the respective government agencies to give little more importance to these. Western Flower Thrips have at least a few hundred host plants range and TSWV can easily infect 100 to 200. So, in addition to *Thrips palmi,* we have a problem on our hand, which needs some serious attention.

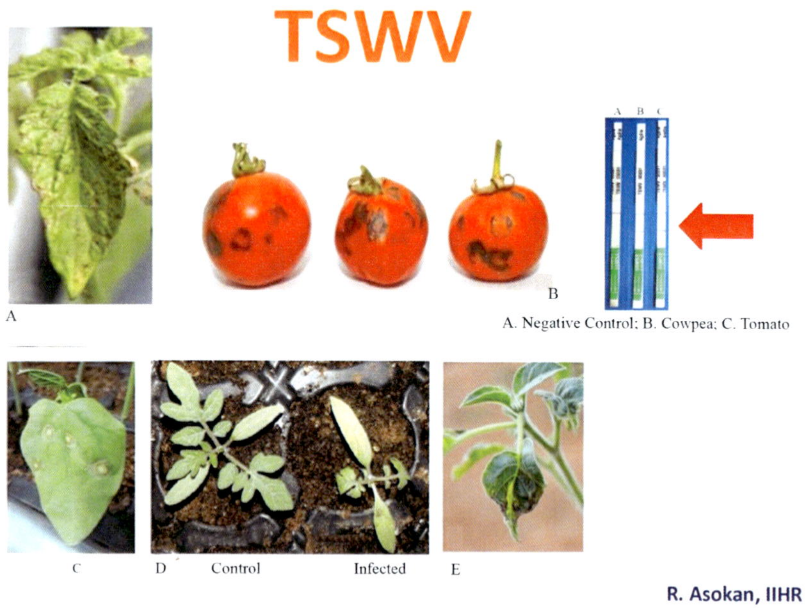

Some studies were carried out at IIHR with regard to TSWV and we can see the classical differences in symptoms between TSWV and GBNV. I would like to appeal both to the private sector and to the public sector that if you have a possible new virus disease, without waiting, please bring it to the notice of the officials. Also, please do not jump into publishing. It is very important to bring it to the notice of the concerned so that the problem is nipped in the bud itself. Early detection of these kind of virus problems can save a lot of money in the future. Now this is with regard to the typical negative cowpea and TSWV classical experiments which carried out by Dr. Asokan. I think if India has to become a developed and scientifically advanced country, we should be open with regard to what we have and what we don't have.

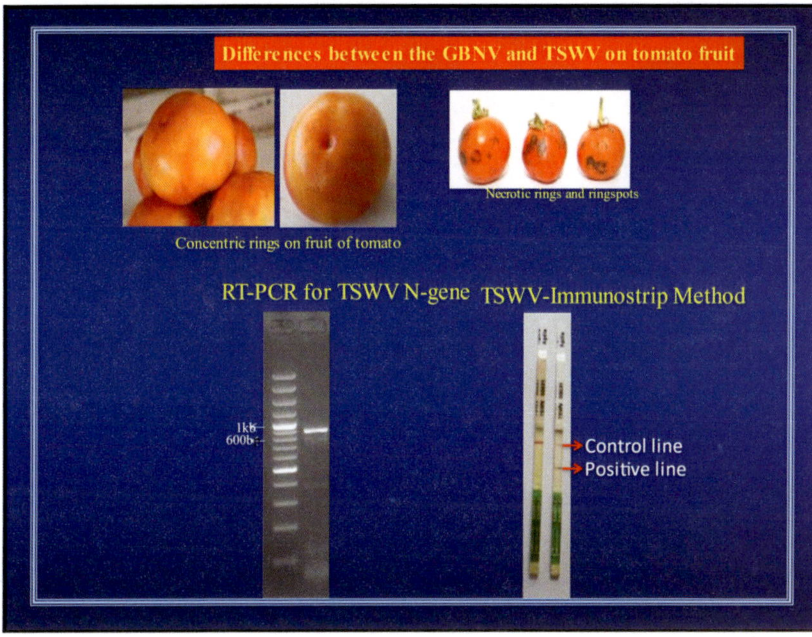

This is with regard to the differences between GBNV and TSWV on tomato for symptoms and identification of the virus. The differences in symptoms are clear and I will not go into its details.

Tomato

GBNV	TSWV
• *T. palmi*	• *F. occidentalis*
• *S. dorsalis*	• *T. tabaci*
• *F. schultzei*	

These are the thrips species associated with GBNV and TSWV in tomato.

Other Crops

Sl No	Crop	Thrips
1	Chrysanthemum	• *F. occidentalis* • *F. schultzei*
2	Capsicum	• *C. claratris*
3	Melons	• *T. palmi*
4	Onion, Garlic...	• *T. tabaci*

These thrips species can also affect Chrysanthemum, Capsicum and many other crops.

We also did transmission studies with regard to GBNV on tomato. We were one of the first groups in IIHR to show the transmission from the acquisition to the transmission stages through GBNV, both in cowpea as well as in tomato.

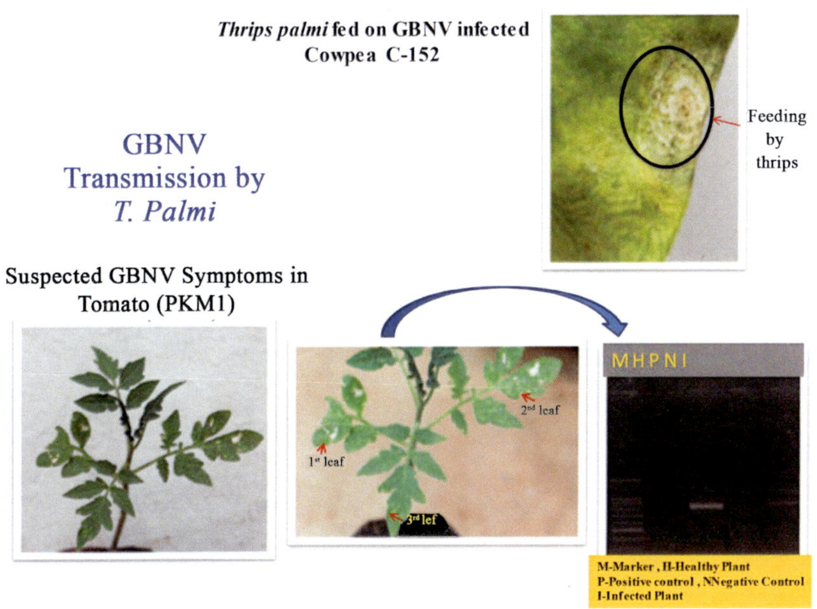

What I would like to ask is - is it possible that there are a few other species of thrips which probably are transmitting about which we know very little?

Validation of species-specific marker

Primer Details:

Name of the insect pest:	Primer sequence	Binding region	PCR product size (bp)
Thrips palmi	KBR (2) F- 5′- TTA AAT TTA CGA ACA TCA ATA AAG T -3′ KBR (2) R- 5′- TAC AGG ATC TCC TCC CCC TCT TG -3′	68-92 613-635	567
Scirtothrips dorsalis	Rebi (2) F- 5′- CTT AAT CTG CGT GTA CCT ATA AAA ATT TT -3′ Rebi (2) R- 5′- TCA AAC AAA TAA AGA TGT TTT ATC TGA A -3′	68-96 481-508	440

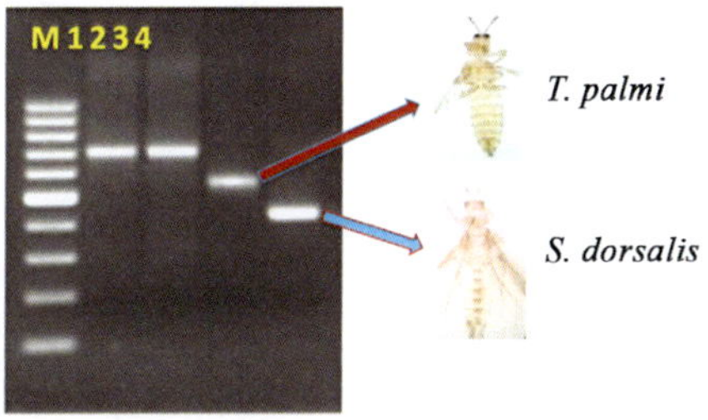

Legend:

M- 100bp ladder, 1&2 PCR amplicon for LCO/HCO *T.palmi & S.dorsalis* respectively, 3 *T. palmi* specific marker, 4 *S.dorsalis* specific marrker.

One of the things that I said in the introduction was about why species-specific molecular markers are important. There are so many species and often they look similar and we don't have good Insect taxonomists for all species of insects. Further, it is very cumbersome to identify them. But through our own work, we developed Molecular systematics for many species of insects and species-specific markers for many sucking pests which in an instant can identify the species.

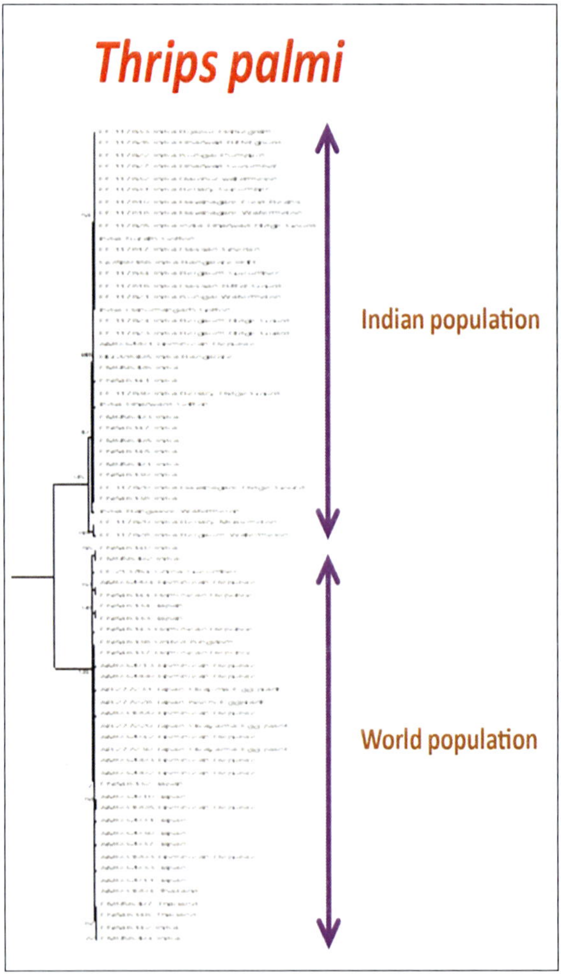

One of my PhD students showed that *Scirtothrips dorsalis* is also a vector of TSWV, but it was a very poor transmitter. When it comes to *Thrips palmi*, there is watermelon silver mottle in Taiwan and in other places there is no Watermelon Bud Necrosis in spite of presence of the vector. Why is it that only

Indian species of *Thrips palmi* are transmitting the Watermelon Bud Necrosis? And, when we compared the sequences of Indian populations and populations from other parts of the world for *Thrips palmi*, we found that the Indian *Thrips palmi* is all together in a different cluster as compared to many other clusters in different parts of the world. Likewise, the Indian population of *Thrips palmi* is the one which has tremendous potential and this is significantly different from the other clusters of the world population. So, there is a need to understand the distribution of the vector and vector status. This is called as vector efficiency. Vector efficiency is, for example, we may have a vector which is an efficient transmitter but their numbers may be very low. We may have a poor vector but their numbers may be very high and their multiplication rate is so high that they may feed on a higher number of polyphagous plants which then becomes an important factor.

Future line of work

- Survey for distribution
- Sensitization on the vector status
- Floriculture movement and Trade
- HPR has to have a relook Eg. Tomato
- Host plants for vector, virus and both
- Other viruses that this can transmit
- Replication and population pattern

In future, we need to take a relook at host plant resistance. I am trying to convey this from a breeding perspective. We have to look factors like, from the perspective of which tospo virus, which vector-transmitted tospo virus, which part of the country the resistance is applicable, etc. There are many other aspects which need to be considered, which include - can we have resistance both for the vector and the virus? The fact is that, very often resistance to the vector does not translate to resistance to the virus. Which other viruses the other thrips vectors can transmit? What is their population pattern? What is the replication pattern of tospo in these species? What is the acquisition and transmission efficiency in that species? And, what proportion of the population can transmit the virus/ host range etc. These are all critical in our management of thrips and thrips transmitted viruses at the field level.

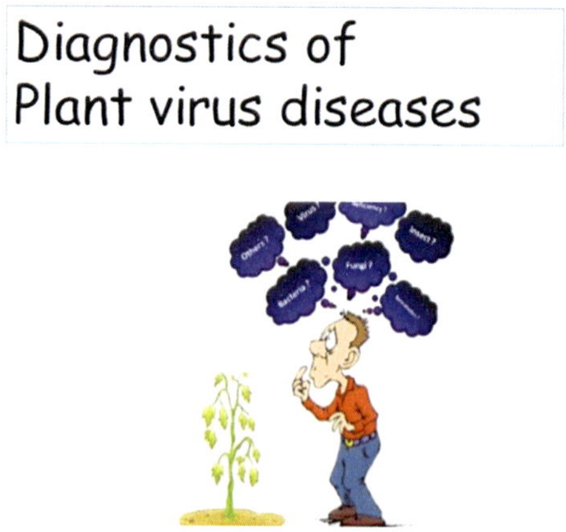

Coming to diagnostics of plant virus diseases, molecular biology has very rapidly advanced. Though initially we were carrying out all kinds of laboratory techniques like indicator plants, sap inoculation, etc., now we have moved pretty far using PCR based diagnostic tools.

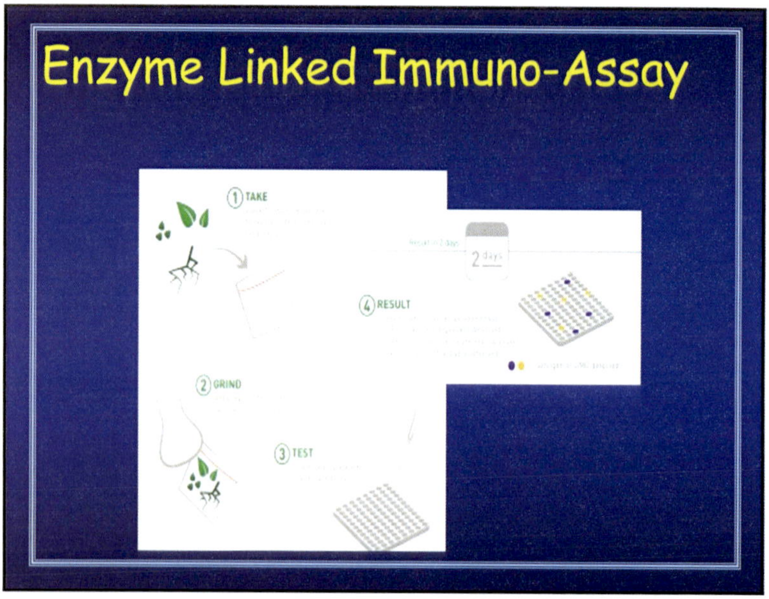

We are now using very sensitive tests like ELISA and other lab tests which can determine the presence of the viruses.

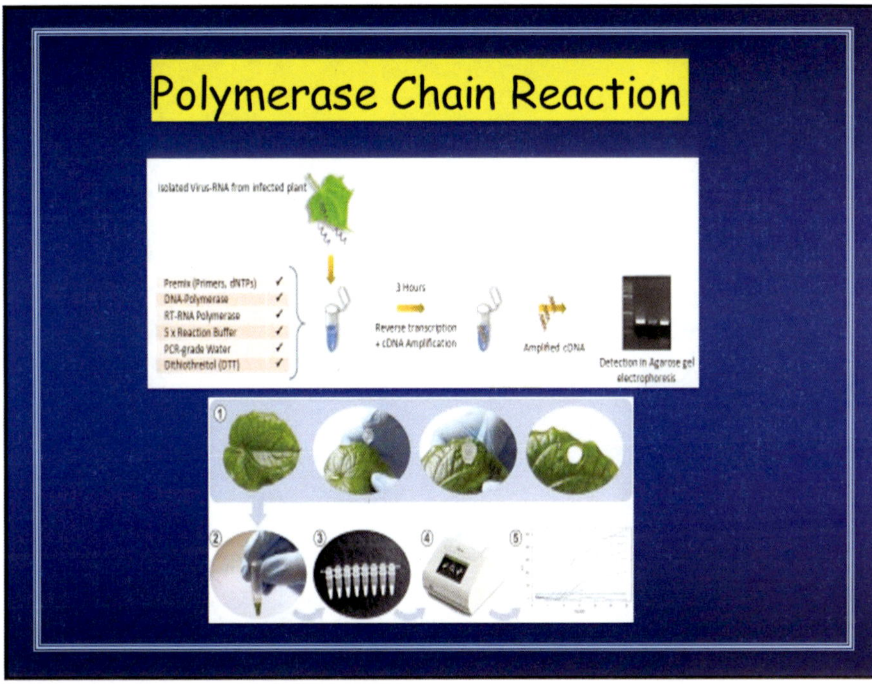

The question of making the final decision in these sensitive tests is the cutoff value. For example, when seedlings are being tested to make a decision that they are free of tospo virus, if the buffer value is 0.1 and if the seedling value is 0.2, are we going to reject or accept it? If the seedling value is 0.3, are we going to say it is healthy or not healthy. These are some of the decisions, like in case of covid testing, which becomes very difficult as to where to draw the cutoff lines. And, for those of us who have to make decisions on how and when to accept imported planting material as free from virus, if the tests show as compared to the buffer even 0.2 or 0.3 as the TSW value in ELISA or PCR, we should reject it. Whereas, in the field screening, if we are getting 0.2 or 0.3, it is better to wait a little longer and confirm before declaring the results. These are tricky scenarios for decision making even with the best of diagnostics.

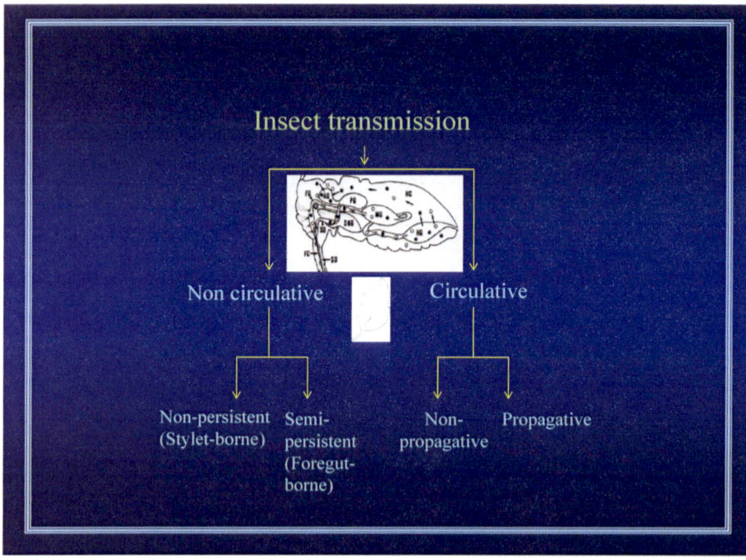

In the case of tospo viruses, they are, circulative and propagative. That means the virion enter the gut, crosses the midgut epithelial barrier and get into the blood or haemolymph. From there, like in mosquito, enter and replicate in the salivary glands. And then adult thrips are ready to transmit the virus.

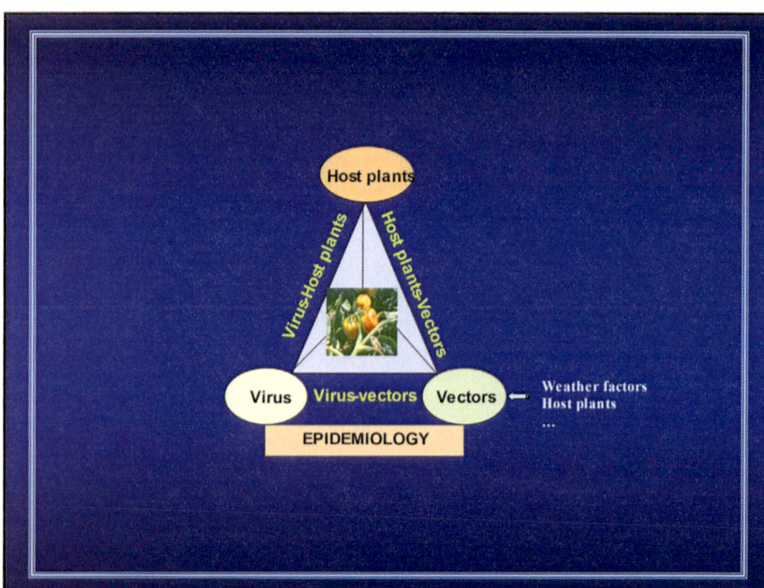

This, I have already explained, is the complex process of epidemiology associated with the thrips-transmitted virus diseases.

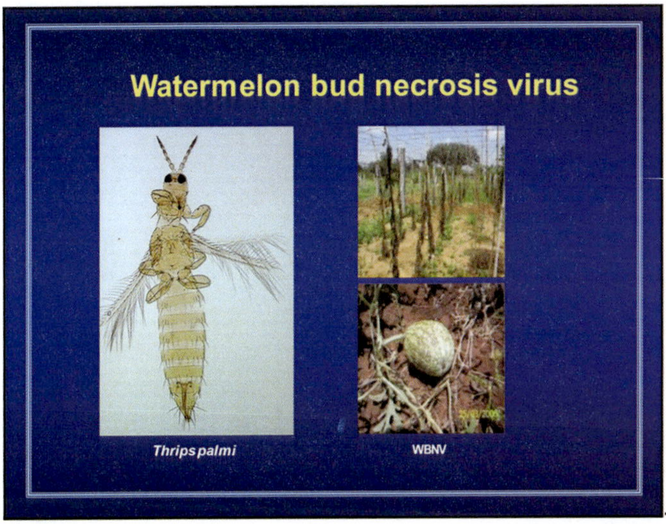

Some of the newly-introduced tospo viruses are highly virulent as we have seen in the case of watermelon bud necrosis. There could be 100 per cent damage to the crop. In Karnataka, this disease is called "jaundice", where from February 15th to April end, if there is good number of *Thrips palmi* present, absolute devastation of the crop can take place.

While we are breeding for resistance to any particular tospo virus, in certain crops, other major virus diseases also have to come into picture, which need to be considered. For example, we cannot lower the guard with regard to TLCV in tomato or other similar diseases that are present.

Another vegetable crop in which we are looking at tospo virus disease is the onion, where in Iris Yellow Spot virus in prevalent. Dr Rawal and I, way back in the 90s, predicted that there is a possibility of Iris Yellow Spot coming into our country. In a matter of 10 to 15 years, it did enter into our country. I am told by the seed industry that only 15 to 20 percent of our onion seed is produced by the organized seed sector, while 80 percent of onion seed is farmer's own seed. Iris Yellow Spot virus is a *Thrips tabasi* transmitted virus. The onion thrips is causing significant losses in certain pockets of Maharashtra. Onion seed is a very critical input. A couple of years back there was a serious shortage of onion seeds and we had to import them. Seeds of the short duration varieties, which are needed in our country, were not available. As a result, we had to import seeds of the long duration varieties from abroad which were no preferred. Onion is a critical segment; onion seed is a critical requirement, and Iris Yellow Spot virus has a critical role in this.

The classical symptom of this virus disease is that it affects the inflorescence which will show diamond shaped lesions and the inflorescence bends and withers. Onion seed crop is a rabi season crop and the thrips on onion will shoot up especially in Maharashtra and Northern belt, from January. So, IRSV needs to be seriously looked into to ensure stable seed supply of onion in our country.

There are TSWV in other crops.

This is Capsicum chlorosis which is another thrips-transmitted virus. Western Flower Thrips is associated with this disease. It was reported also in India by the Mahyco group long back. However, not much information has come out after that.

When we are managing thrips in the open fields as compared to under protected cultivation, we need to be aware that thrips are a notoriously serious problem in protected cultivation. Whether the crops are grown in protected net or open, we need to have different management strategies. Resistance in the crop to thrips in polyhouses is way higher as compared to open field cultivation. We cannot assume that the high levels of resistance observed in polyhouses will be observed in crops grown in open field.

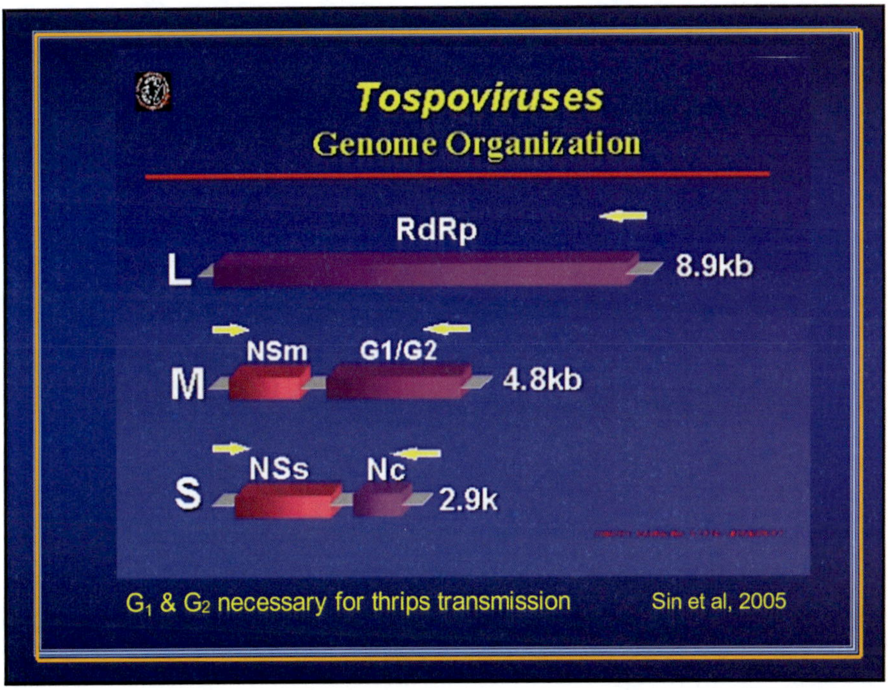

Now a little bit about the genome organization of the tospo virus. It is a tripartite virus. There are large, medium and small segments RNA. RNA dependent RNA polymerase (rdrp) is the enzyme which starts the life in itself. This is coded by the large-segment RNA and the medium segment RNA codes for the two spike proteins. We call them G1 and G2 in tospo virus because it is a single protein which is cleaved into two. That which is near the carboxylase end is called the G_C, and the one which is near the amine end is called the G_N. It also codes for one non-structural protein. This small segment RNA actually codes for the nucleocapsid protein, i.e., the coat protein, and also it codes for the non-structural protein. Non-structural proteins are those proteins which are produced by the virus when it is causing infection. For example, when there is dengue, we look for the NS protein which means that the virus is replicating and the non-structural proteins are produced, which acts as a defence to overcome resistance from the host. So, if we are breeding for resistance, an interaction between the resistance factors operating in the host plants, and how the non-structural proteins associated are being produced, and the regions coding for them in the virus become most critical if resistance has to be durable.

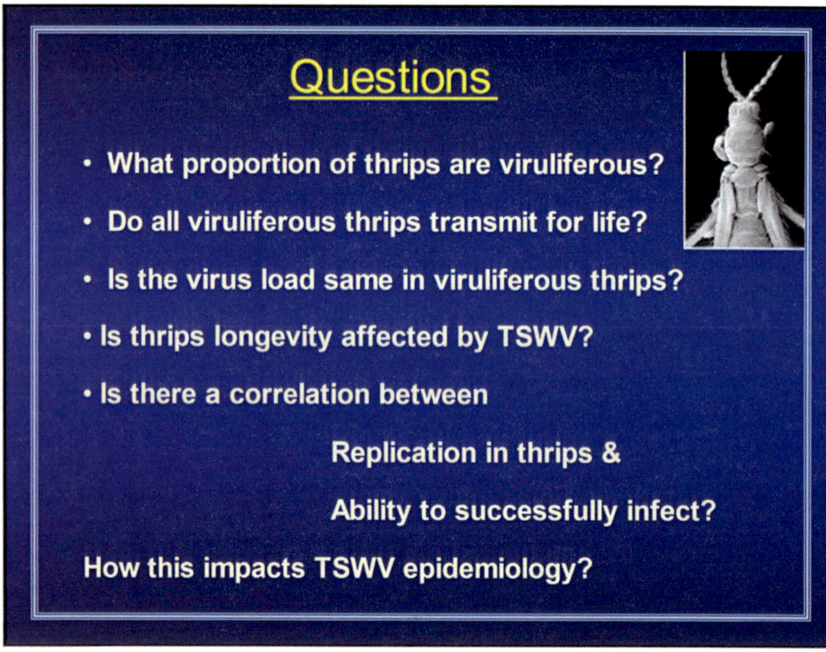

So, in breeding for resistance to the viruses, the key questions to ask are - what proportion of thrips are viruliferous in different species? Do all the thrips transmit for life? Is the virus load same in all the thrips species? Is the longevity of thrips itself affected by TSWV? Like in the case of fleas, where they get affected by the plague bacillus and not able to survive for long, does this hypothesis hold good which may impact the life of the thrips? Is there a correlation between replication in the thrips and their ability to infect across thrips vectors? These are some points which will impact the epidemiology, and these were also some of the questions which I tried to answer in my postdoc work. When it comes to tomato spotted wilt virus, unlike many other aphid-transmitted viruses, repeated mechanical inoculation of the germplasm causes defective virus particles which can cause the infection but will not be acquired by the thrips anymore. That means, the virus particle which is being used for sap inoculation repeatedly can actually be a defective TSWV particle which has less relevance to what is happening in the field. In the USA, what they do is that in one cycle, they go through mechanical inoculation and in the second cycle they go through thrips inoculation. Alternatively, we may infect some of the perennial plants like *Datura*, *Emilia vastatrix*, eggplant etc., where we can repeatedly harvest young leaves to get the thrips-transmissible isolate for any of the screening program. This is very critical for those who are carrying out germplasm screening for tospo virus resistance.

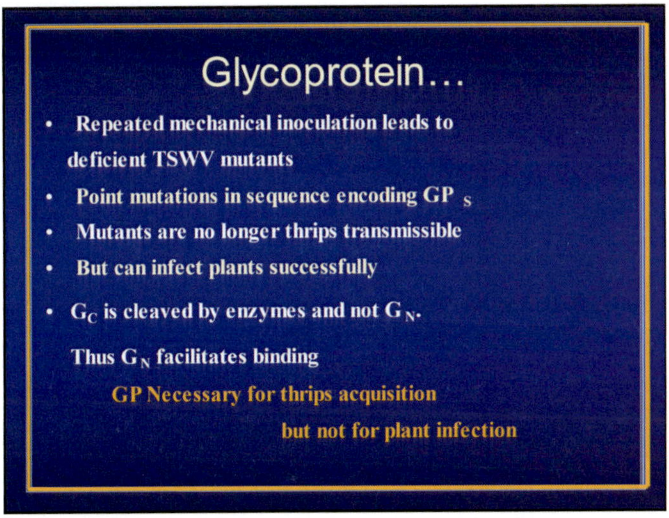

Point mutations occur in the glycoproteins after repeated mechanical inoculation which makes it thrips non-transmissible. But they can infect the plant through mechanical inoculation because G1 and G2 are not needed for infection.

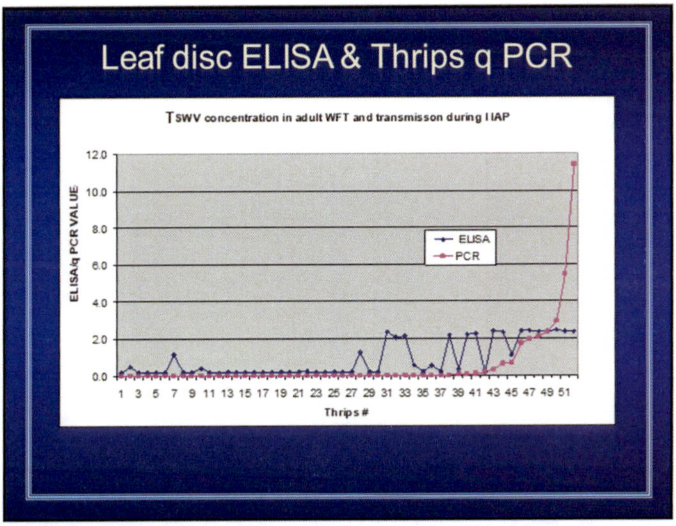

This is part of the work that we carried out with regard to the rate of TSWV titre in the WFT and transmission in the first three inoculation after the adult emergence. The $_Q$PCR values were high in thrips transmitting the virus.

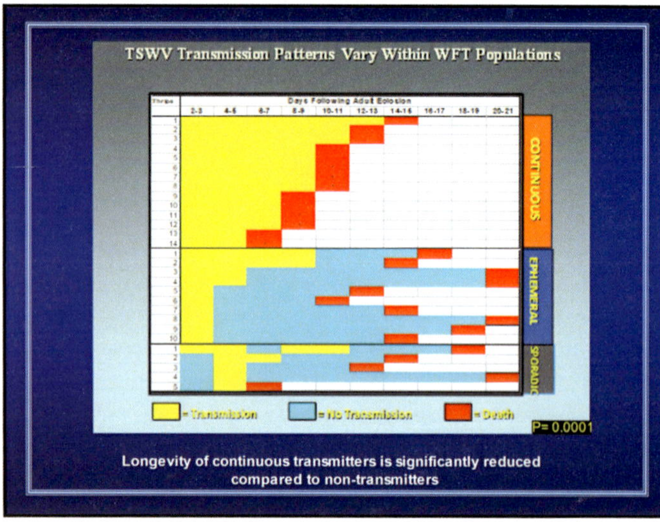

We have observed that there are continuous transmitters and ephemeral transmitters. Continuous transmitters can transmit for a period of at least 15 days. These individual transmission studies were carried out by me in UC, Davis with not less than 50 to 100 individual thrips, day after day. The ephemeral transmitters, at most, can transmit when provided 2-3 inoculation access. The sporadic transmitters transmit once or twice in their life time.

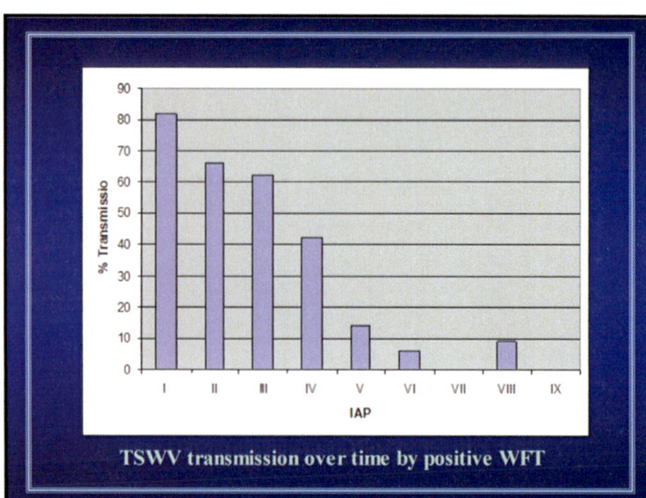

During screening studies, it is possible that sometimes using a thrips colony which is successfully transmitting the virus, we may actually get very poor transmission. Our studies clearly indicated that as soon as the adults emerge, within one or two days of their emergence, the transmission is most efficient. Third and fourth day, it is little less. Fifth and sixth day, it is even lesser. And after some days, there may be no transmission. So, if we are doing thrips transmission studies, it is best to make sure that immediately after adult emerges, use the fresh adults, which are very active, for transmission studies, especially in the final germplasm screening programs. This will ensure that screening tests will be highly successful and reliable.

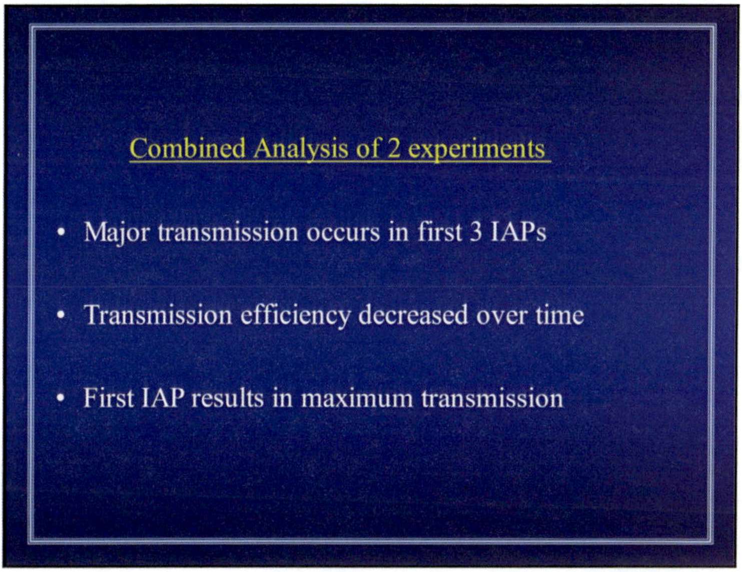

So, the major transmission occurs in the first three inoculation access which means that as soon as they emerge the first three inoculation access that are given, they will be very efficient. It decreases over time. The first inoculation access is the most efficient.

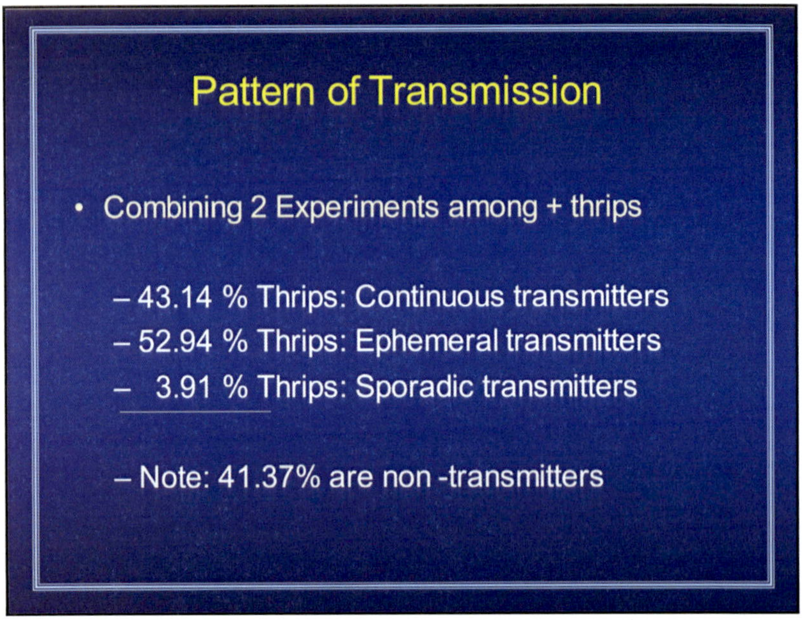

The combined analysis of the two experiments indicated that the ephemeral transmitters form the predominant type, followed by continuous transmitters and sporadic transmitters are minimal.

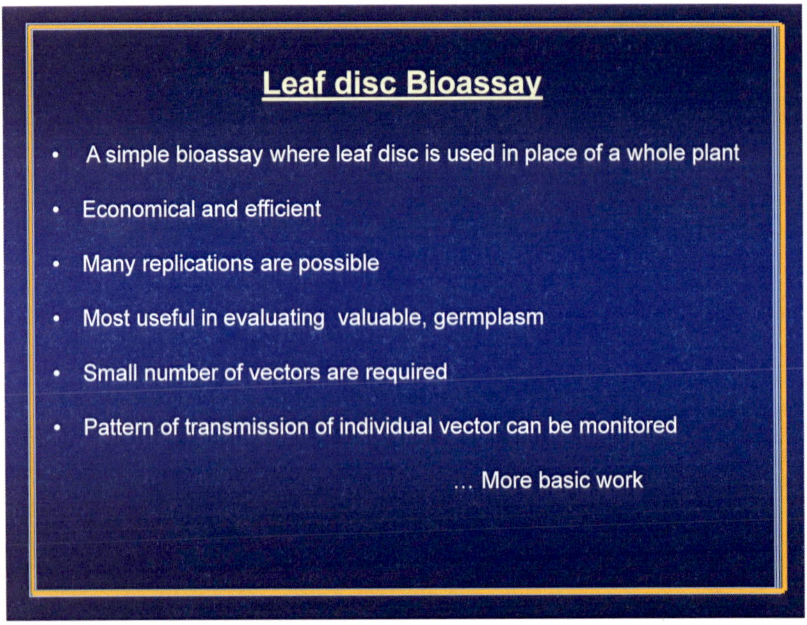

We developed an assay to screen large numbers of valuable germplasm lines.

We developed a leaf-disk bioassay to screen by cutting out a small leaf disk, the size of a one rupee or a two-rupee coin, and putting them into a long glass tube and allowing the adult thrips that are freshly emerged which have acquired the virus to feed on the leaf disc in the glass tube for about 24 hours. After 24 hours, the leaf disc is removed from the glass tube using a long forceps. The disc is then floated in a Petri dish containing water. This will help the leaf disc to regain its turgidity. After about three or four days, leaf disc is analysed using ELISA or PCR which indicates whether it is TSWV plus or minus. We can do 50 replications from the same plant or different plants. This test helps in saving the plant and also determining whether it is susceptible or not. These tests can be carried out with any kind of plants, e.g., *Lycopersicum chilensi* or *Lycopersicum pinnellii*. Using these kinds of leaf disc bioassays in controlled experiments in the laboratories, we can determine resistance in any germplasm line.

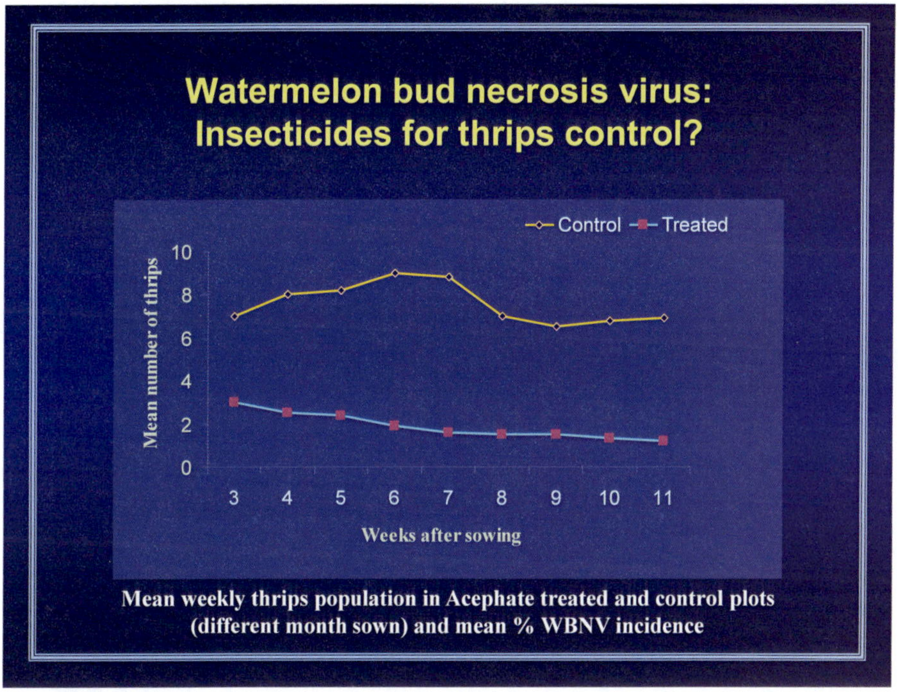

Mean weekly thrips population in Acephate treated and control plots (different month sown) and mean % WBNV incidence

This is another experiment to show whether insecticides can control thrips which was carried out in a field where we had a serious incidence of watermelon bud necrosis virus. Treated plots showed less thrips population as compared to the unsprayed plots.

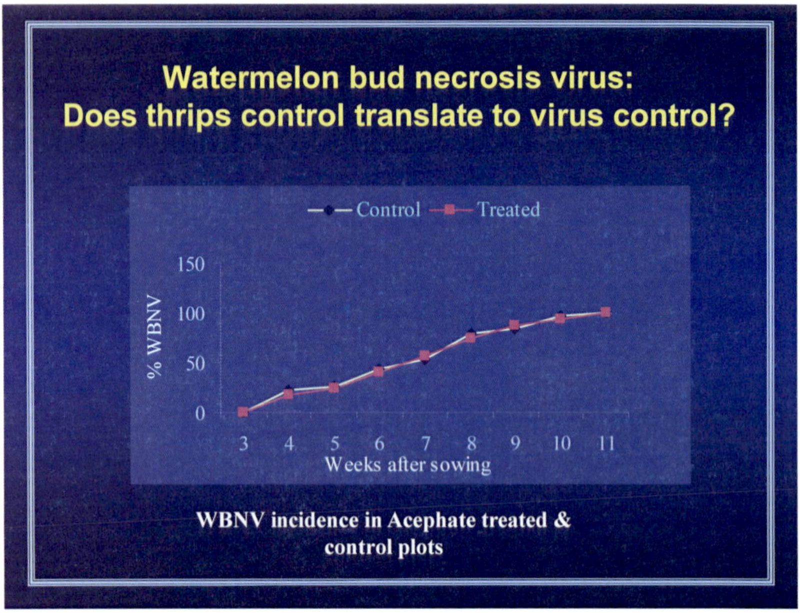

But, as we can see, between control and treated, there was absolutely no difference with regard to the watermelon bud necrosis incidence. It shows very clearly that insecticides are not the answer to management of bud necrosis on watermelon.

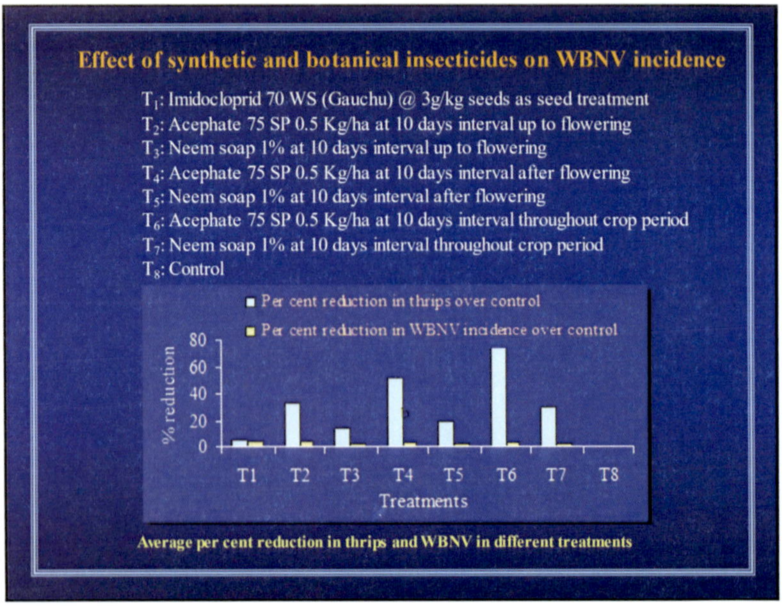

In a field scenario, we have millions of thrips which are moving around the field. It takes only one or two probes for them to transfer the disease. Even if the thrips die, if they are able to transmit the virus before they die, the virus will be there and we will not be able to control. So, these are the kinds of problems that we encounter in management of the thrips-transmitted virus diseases. That is why in such a scenario, host plant resistance offers economically and ecologically the most viable alternative to chemical control. I am told that in IIHR already advanced stages of resistance are available for the Watermelon Bud Necrosis Virus, but it is yet to be confirmed. We found that there is hardly any reduction in Bud Necrosis Virus between different chemical treatments which shows that insecticides usage is futile. So, the only way is to go for genetic resistance.

When I was working in Bioversity International, there was a strong program on conservation and evaluation of crop wild relatives (CWRs). This, I believe, is something which we have to go in a big way. Even I have been suggesting to NBPGR to give priority to genotype conservation of the crop wild relatives because it is the CWRs that are really going to contribute to resistances in the future.

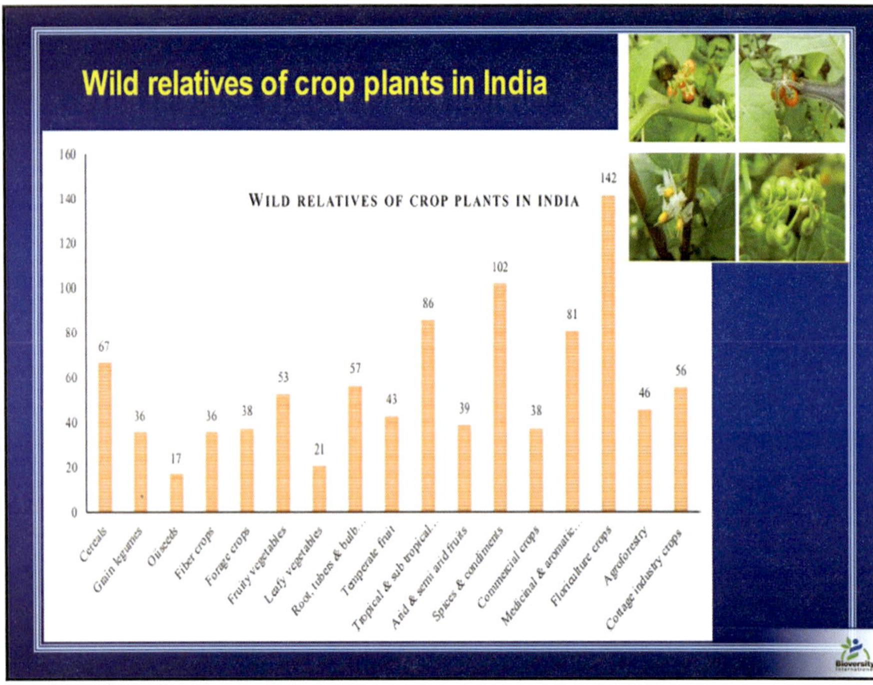

Whether it is okra or tomato or maize or any other crop, the genes that have come from the wild have made a tremendous difference to our life. The chart indicates the wild relatives which are recorded in different crops. It indicates the different numbers of wild relatives that we have. We can also see that the maximum number of wild relatives are in floriculture crops, spices and condiments, tropical and subtropical crops. We do not have figures for leafy vegetables, fruits & vegetables in this chart. But it is not how many wild relatives we have; it is how best we are able to make use of the crop wide relatives in breeding.

One of the big advantages in horticulture crops is that the crop wild relatives can also contribute significantly as rootstock in addition to their utility in the breeding programs. Crop wild relatives can confer resistances to a number of factors. For example, the crop wild relative which we may be using for nematode resistance may also contribute to moisture stress tolerance or resistances to soil-borne pathogens. This slide shows *Solanum chilense* being collected. *Solanum cheesmaniae,* collected from another region, has contributed to salinity tolerance. We may have to go to different parts of the world to collect them. In India itself, in many crops like cucurbits, okra, etc., there are excellent collections available. While we have made good germplasm collections, we have not evaluated and profiled them in a scientific/systematic way for abiotic and antibiotic tolerances. For example, if we want to select a particular genotype for salinity tolerance, or for some other trait, we should have such information instantly showing that this is the germplasm that we can use in breeding/grafting. However, since that kind of a research is basic research, it is absolutely essential to be carried out. The pesticide and seed companies may also look into utilization of the crop wild relatives to add new molecules in pest management.

We need to look at CWR both from a climate change perspective as well as from disease resistance perspective. We have to shift the goal post in utilizing crop wild relatives. That means we have to broaden our horizon and change the way we are looking at these valuable inputs that are available for us.

CWR Breeding – is Difficult but Possible

- Undesirable traits in wild relatives- "Linkage Drag"
- Wild plants are difficult to manage under "domesticated"
- Crossing barriers
- Long time requirement
- Protocols are needed for screening
- Crop wild relatives accessions are much less represented
- Access to germplasm (CWR, but also crops) collections
- Information on already existing pre-bred materials is often missing or unavailable
- Often few available evaluation data on crop wild relatives
- The taxonomy of crop wild relatives is sometimes confusing

Introgression of genes from crop wild relatives is difficult but not impossible. There are linkage drags during the breeding process with crop wild relatives, but molecular biology can make a difference. Also, plants of crop wild relatives are difficult to manage under domesticated conditions. The extent of deforestation that is taking place is something which bothers me. With increased urbanization, before long, we may lose many wild relatives that exist now. So, there is an urgent need to collect these wild relatives before they become extinct. While using them in breeding, there are problems associated with crossing barriers. There may also be a need to do embryo rescue to avoid abortion of embryos, which also may not always be possible. It is a long-drawn process but I hope that molecular biology can reduce the process time. There are techniques like doubled-haploid breeding which can reduce breeding cycle or time. Often, sharing of the accession is difficult. And the pre-breeding material is missing and becomes unavailable. This is very important. For example, a lot of hard work was carried out in the field for 10 to 15 years in the development of okra yellow vein mosaic resistance. We observed that the resistance has broken down after 10-15 years. Had we conserved all the segregating populations generated in the process of their development, we could have gone back and reselected for more stable tolerance levels in the segregating material. There is also very little evaluation data. Taxonomy, I was told, by Dr. K V Bhat (NBPGR), that sometimes identification of a particular species is actually inaccurate. This requires a trained botanist to say what species it is.

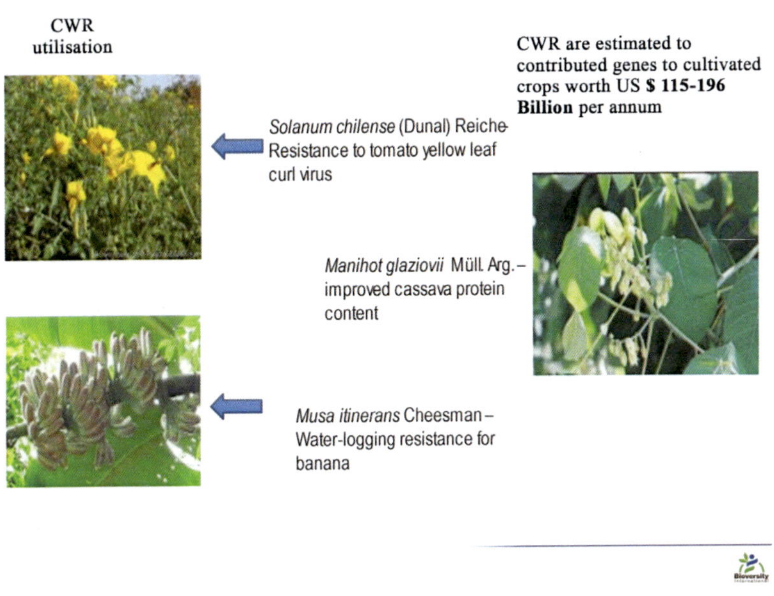

I will not go into the details of this slide, except to say that crop wild relatives are estimated to contribute about 200 billion US$ per annum to cultivated crops in India. It will be much more if we can mainstream them into our agriculture.

This is a classic case of germplasm evaluation. This is my own experiment. On the left side we can see *Lycopersicum pinnellii*. Now the genus name is changed to *Solanum*. We inoculated *L. pinnellii* three times using infected sap and they did not show the symptoms. Therefore, I categorized it as resistant. Then Dr John Cho, who developed the sw5 genes, suggested to again test it with three more inoculations. When I completed six inoculations, *pinnellii* became susceptible to TSWV, whereas thrips transmission on the right showed resistance. Now what happens many a times with our repeated sap inoculations with million times more the dose is we may conclude, in the laboratory, an otherwise very useful field resistance as susceptible. So, these kinds of balancing with reference to nature has to be considered before final conclusions are made. I still feel *Lycopersicum pinnellii* has got very useful resistance to TSWV which, in all probability, also has high level of field resistance. However, if we go with repeated mechanical inoculation, it might show susceptibility like in this particular case.

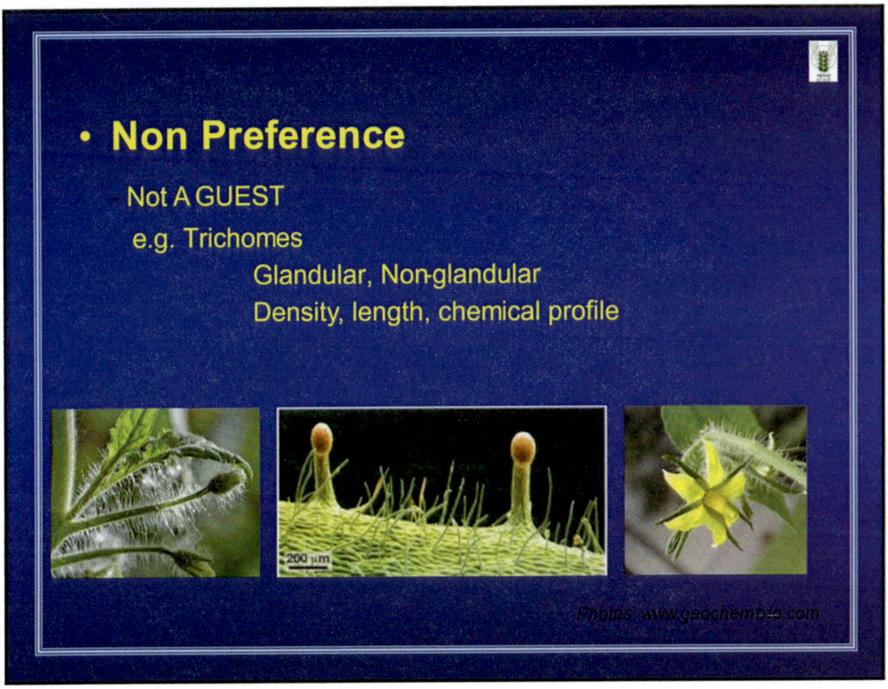

Now about the mechanisms of resistance, we had a high-level group meeting on the future of biotechnology with Dr Swaminathan and the ICAR group. Number one on the list was breeding for insect resistance because traditionally breeding for insect resistance is extremely difficult. As we can see, trichomes are both glandular and non-glandular. There are different types of trichomes. For those who would like to understand the role of trichomes in insect resistance, we may study the works of George Kennedy. Trichomes are the defence mechanisms that every plant has with regard to their own resistance. For example, *L. habrochaites* is one which has highly dense trichomes. The chemical composition, the acyl sugars, and the length of the trichomes, contribute to insect resistance.

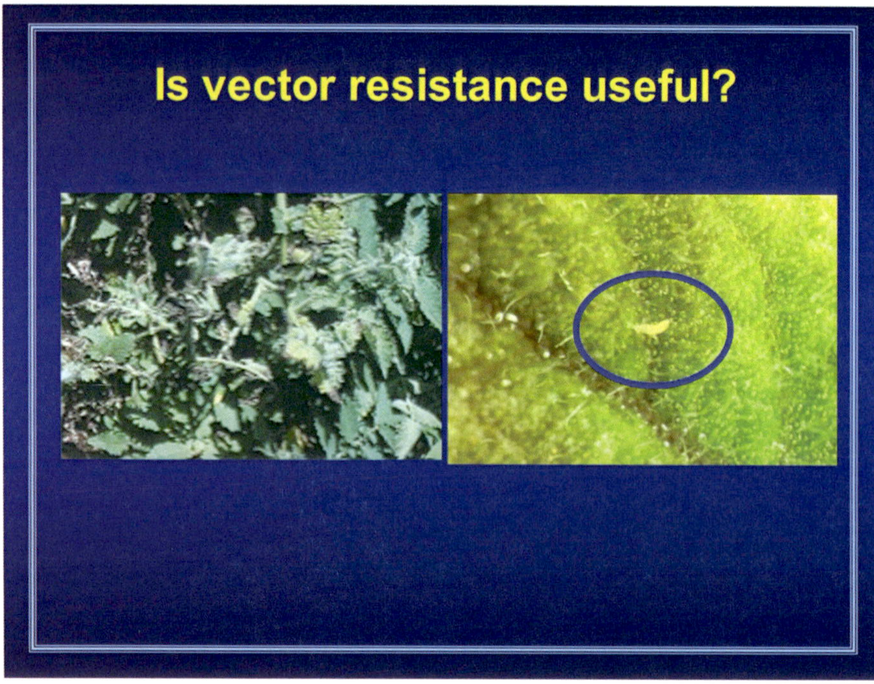

On the left-hand side of this slide is *Lycopersicum habrochaites.* We may be able to see the small blackish spots on the leaves, which may not be very clear. It is field susceptibility to TSWV in *Lycophersicum habrochaites*. Whereas, on the right side of the slide, we can see small bubble-like structures which are glandular trichomes which trap the thrips. When we looked in detail, we found that as senescence was taking place, the plant was losing trichomes, and we suspected that, probably, some thrips feeding was happening on the more mature leaves which had less trichomes density and glandular trichomes. Especially, 2-tridecanol, the chemical which is present in glandular trichomes, though confirms very high level of resistance to thrips, it does not translate to field resistance to TSWV. These are some of the things that need to be very critically looked into because this particular species is extensively utilized for the tomato leaf curl virus resistance. If someone assumes that the same thing will work for tospo, maybe, we are not going in the right direction.

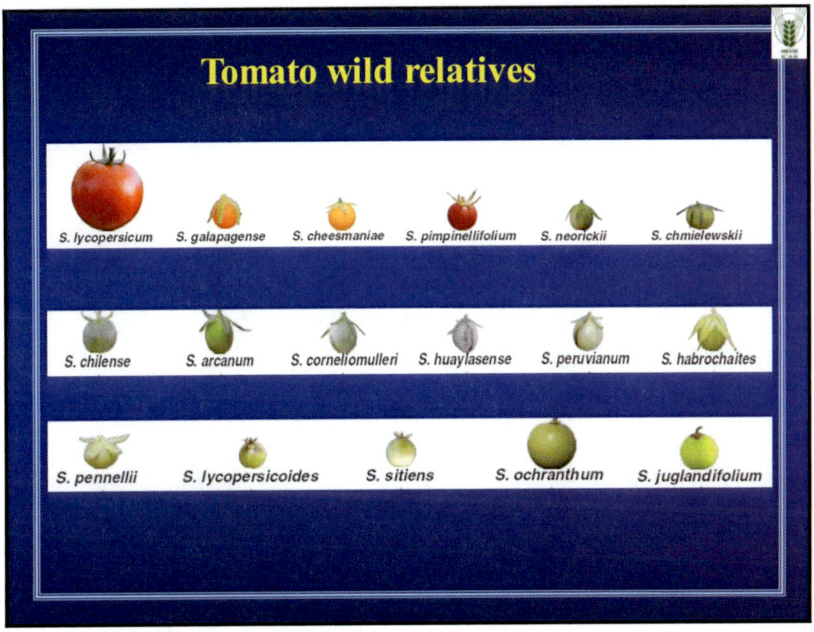

In case of tomato there are many wild relatives. My bet is *Lycopersicum chilense* which we have categorized, based on our work, as one among the best sources for breeding for tospo virus resistance.

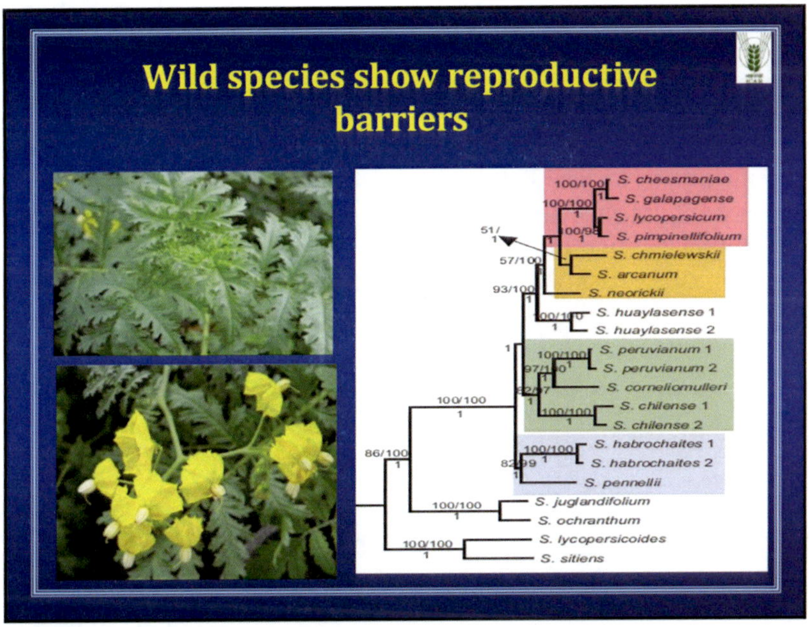

Dr Prasanna, at IIHR, has carried out certain inter-specific crosses between wild species. I had suggested to Dr Prasanna, that in addition to TLCV, parallelly to evaluate the segregating lines for TSWV or tospo virus resistance. However, if the segregating material is still available, if they can be evaluated for tospo, I think, we would have moved a few blocks further in the process.

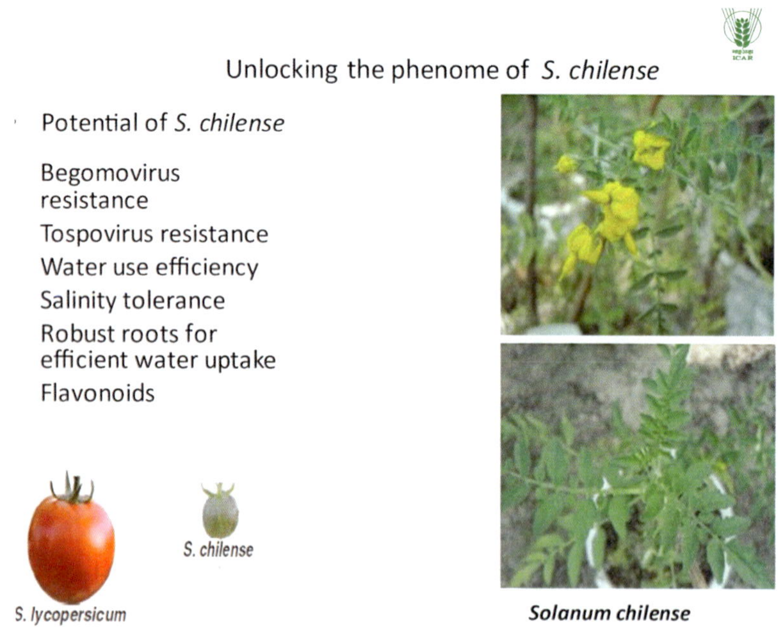

This is *Lycopersicum chilense*, which I feel, both with respect to thrips resistance and other potential beneficial traits, is a very useful source. We did tests both with mechanical inoculation as well as through thrips transmission. The paper was published in 1993 or so, and we did a comparison of mechanical and thrips transmission of TSWV. This is the species we found to be useful and recommended it as the best. In the changing scenario, I would request the tomato breeders to again look into the crossability as well as resistances for advancing this further in breeding. I am not sure what its contribution of resistance to TLCV is, but these are some of the things that need to be taken further.

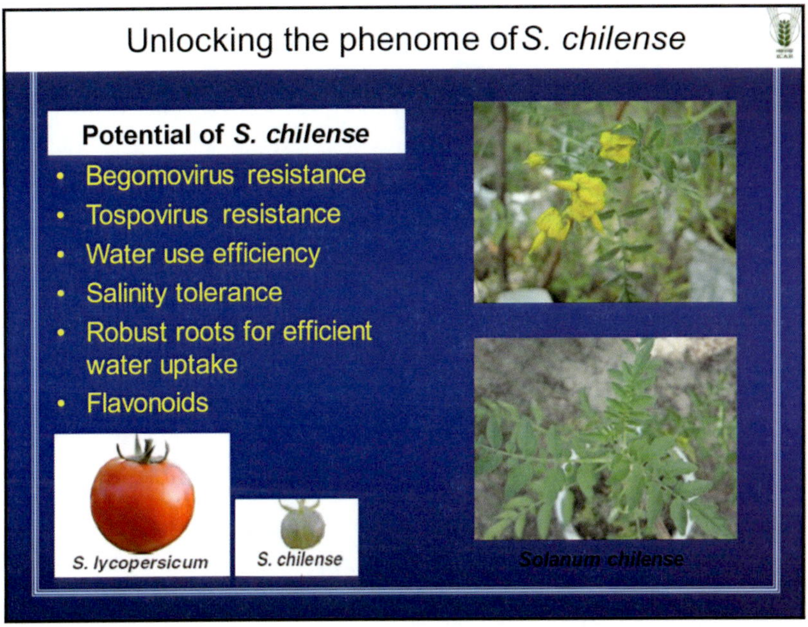

There is potential for *Lycopersicum chilense* in breeding. It also has begomo and tospo virus resistances and provides a number of other advantages in breeding. It is probably worth investing a little more money into this work.

Tomato leaf curl virus

What has surprised me is that in spite of the fact that we have certain amount of diversity in a particular species, whether it is *Lycopersicum habrochaites* or *L. chilense*, it will not be enough if we have only few accessions of these species to work with. We should have a number of collections within the species from different regions to be evaluated.

Proportion of genetic variation in cultivated tomato and their crossable wild relatives (measured by DNA polymorphism)

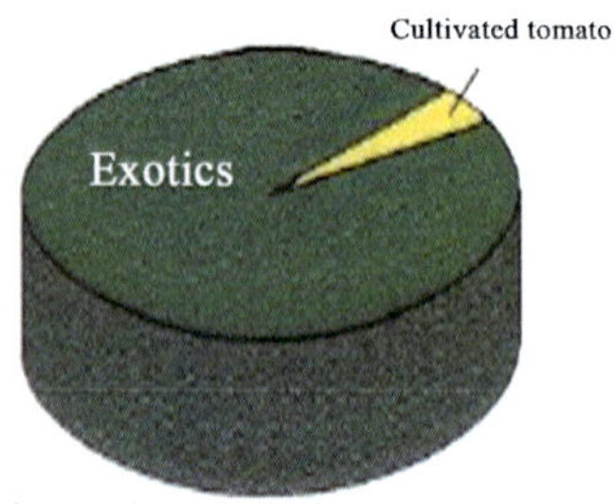

Tanksley & McCouch1997 Science

Tomato an autogamous species in which germplasm is severely reduced by biotic pressure

I personally feel that the future investments and future hope for our sustainable breeding is through utilization of crop wild relatives and we have to go significantly in this particular direction if we have to make future progress.

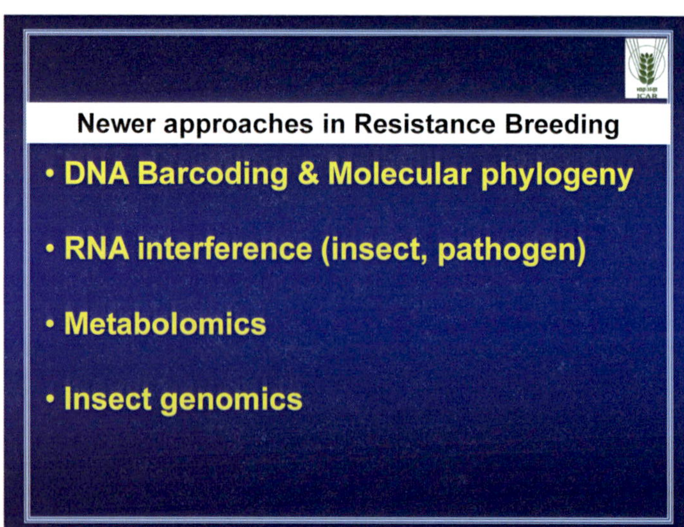

The other aspect is to invest a lot more with regard to molecular barcoding of vectors because half of the times the crop species that we are working with, the vector species are not properly identified. Also, the moment we see a white fly, we should not immediately say it is *B. tabaci* and go in for a spray.

I have a great belief that RNAi interference or gene silencing is one of the most promising technologies, which helps in understanding the in-depth molecular biology of the virus and the insect vector. If we have to move forward for virus resistance or resistance management to insecticides or with other aspects of the plant genome - virus genome interactions, etc., we need to use these technologies extensively. The first and foremost aspect in a vector transmitted virus disease resistance is that we have to look for virus resistance. Virus resistance can be of two types – resistance either for virus replication and/or for virus movement. If there is replication resistance, this is the best option. In addition to that, if there is a resistance for virus movement, it is desirable. Though intensive replication and resistance to movement can often induce what is termed hypersensitive reaction. For example, in case of sw5, there was no systemic movement of the tospo virus, but on the tomato fruits it showed huge blotches which ultimately made sw5 not so desirable for TSWV resistance.

This is the work with regard to gene silencing that has been going on at IIHR and I am very happy that some of the results that I have been able to show here are very promising. The futuristic breeding processes have to combine a lot of molecular biology along with the introgression from the wild genes if we have to go in a big way. And as far as I see, for many of the problems if we are looking within the cultivated species, it may not be able to provide the durable resistance that we may desire.

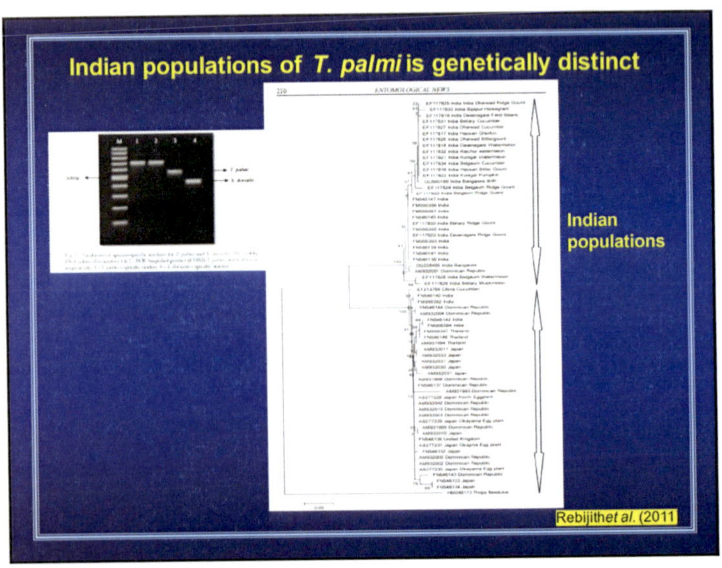

I will not go into this slide as I was not a part of this work.

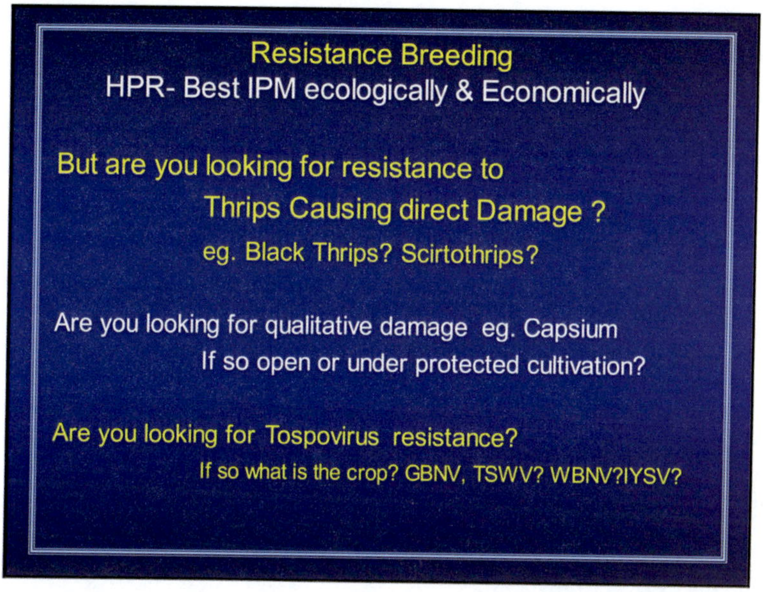

Resistance breeding, i.e., host plant resistance, offers the best IPM component ecologically and economically. Now I want to summarize this for thrips resistance. I would like to ask the plant breeders and others involved in host plant resistance - are you looking for resistance to thrips causing direct damage like in the case of black thrips or *Scirtothrips dorsalis* in chilli, or are you looking for qualitative damage like in the case of capsicum or pomegranate or grapes or cardamom, etc., where the qualitative damage is something which is of a higher concern, or are you looking for tospo virus resistance? If it is for tospo virus resistance, then there are many tospo viruses – GBMV, TSWV, WBNV, etc. I am told that WBNV is also infecting Solanaceous crops. In case of IYSV, it is infecting the seed crop. So, if we take Solanaceous crops for e.g., chilli, capsicum, paprikas or tomatoes, both grown outdoor or indoor, GBNV, TSWV or WBNV are important in addition to direct leaf curl caused by thrips. So, there is a bouquet of problems that are emerging and unless we are sure only then the traits can be suggested to breeders. For example, if we are looking for direct damage, as in the case of chilli, what we have noticed is that the resistance to leaf damage, i.e., if the leaf curl symptoms, can be significantly reduced, we can prevent a lot of direct damage due to *Scirtothrips dorsalis*. But if we are looking for resistance to black thrips, we should only focus on those genotypes which are able to resist black thrips post monsoon when the pest is a problem and *Scirtothrips* is not a problem. But, if we are looking for resistance in capsicum, cardamom or any other crop, what we have seen is that thrips are not negatively phototropic but they are thigmotactic and would like to hide in crevices. Most of the thrips, except black thrips, would like to hide from the rain. So, in the case of capsicum or in certain chillies, we see that post-fertilization, a part of the flower petal or other debris sticks to surface of small developing fruit and the thrips would love to be there at that particular stage hiding underneath the developing ovary or developing fruit. They scar the fruit at a young stage, which when the fruit develops, result in scaring damage. This is what happens in case of grape thrips too when it is at the cumin stage. Same thing happens in pomegranate and cardamom where it hides in the sheath and if the sheaths are slightly open there will be less of scaring damage. We identified a resistant line in cardamom wherein significant thrips damage to quality of the capsule was prevented. Now when you are looking for tospo virus resistance it's a different ballgame.

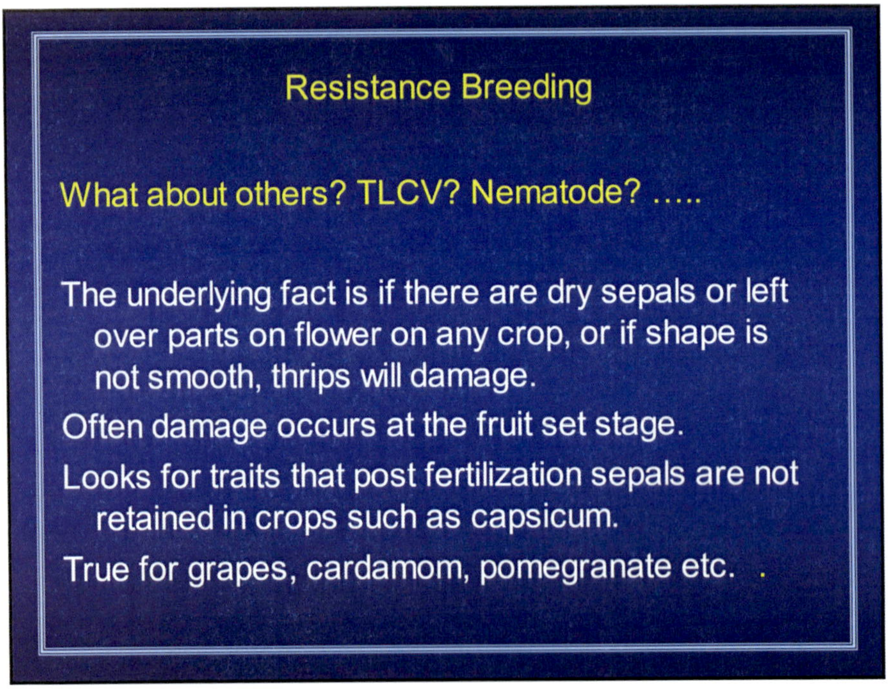

The underlying fact is that if there are dry sepals or leftover flower parts or if the shape is not smooth and especially if the shape of the capsicum fruit is not smooth and it has got too many undulations and too many crevices like in case of California Wonder, the thrips damage will be high. Often the damage occurs at fruit set stage but what we see is the damage at the later stage. Therefore, the breeders should look for traits that express during the post-fertilization stage, like the sepals are not retained in crops such as capsicum. It is also true for other crops too.

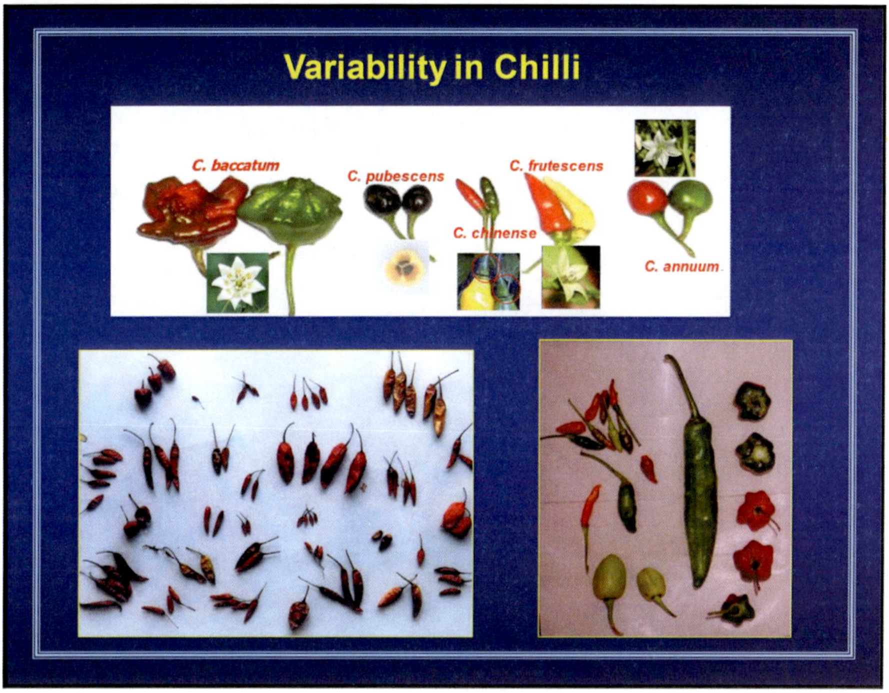

We have innumerable number of wild species in chillies and I don't think a very critical analysis has been carried out with regard to both the black thrips and *Scirtothrips dorsalis*. This needs to be carried out in addition to other aspects.

Capsicum annuum
White corolla & white filaments

C. frutescens
Greenish corolla and
purple filaments

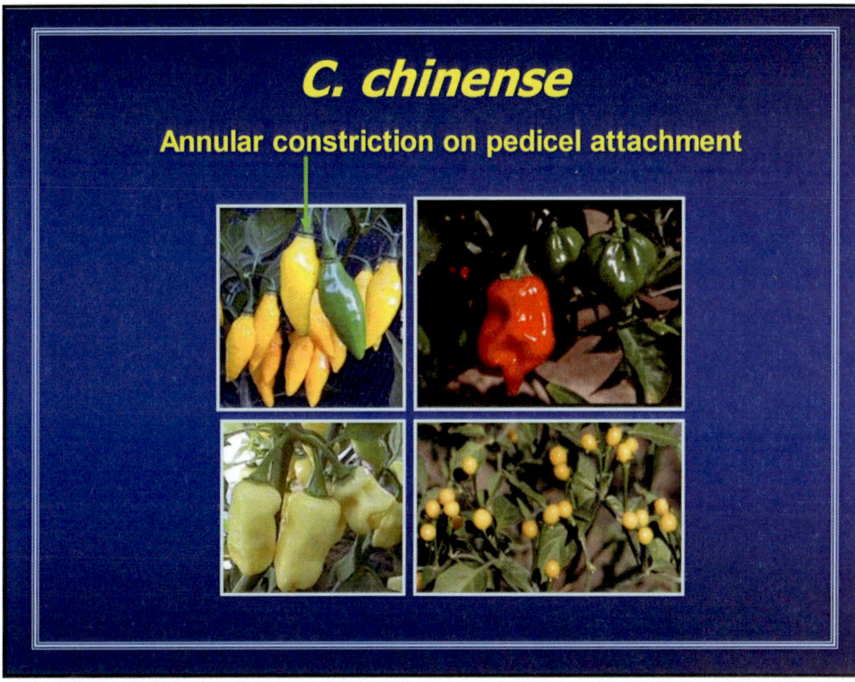
C. chinense
Annular constriction on pedicel attachment

C. baccatum
Yellow spots on corolla

This is with regard to *Capsicum annuum, Capsicum frutescens, Capsicum baccatum* etc. I have a tremendous hope that one/many of these species will contribute significantly in breeding for resistance. I also think that *Capsicum pubescens* has tremendous potential as a source of resistance to thrips.

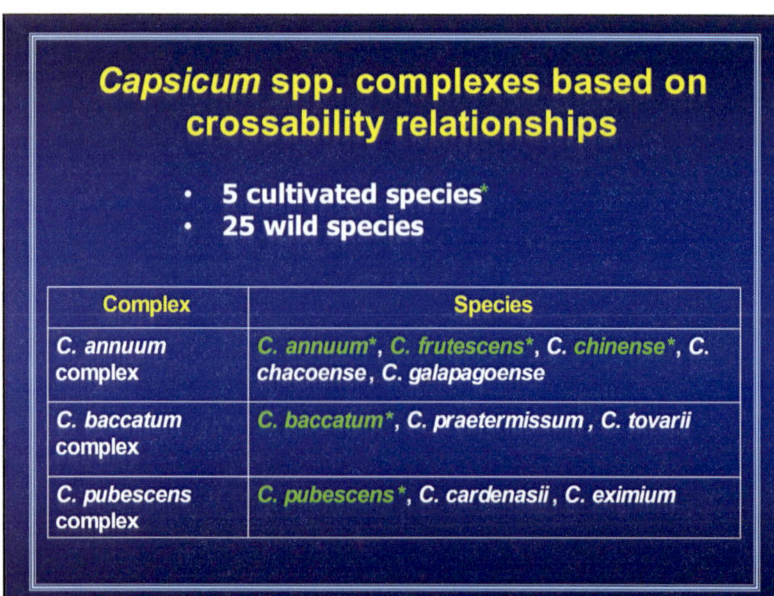

Most of you know that there are different species in *Capsicum* genus which form several species complexes based on their crossability. I will not go into the details.

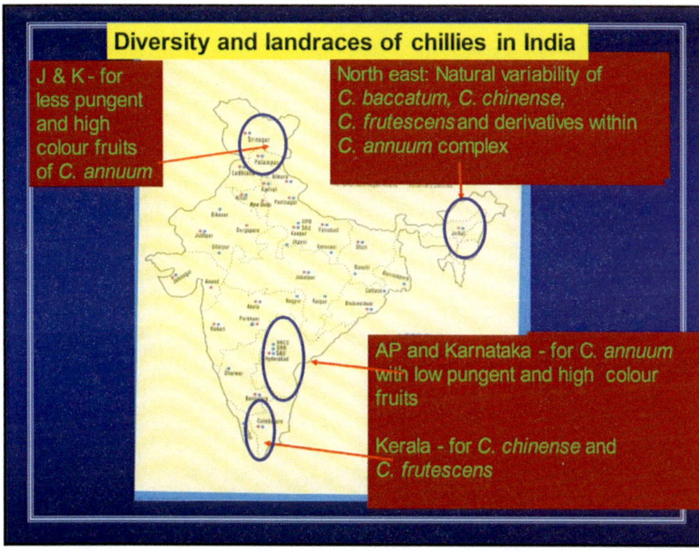

There is tremendous diversity in chillies though India is a secondary centre of domestication. This tremendous diversity is what needs to be exploited in breeding.

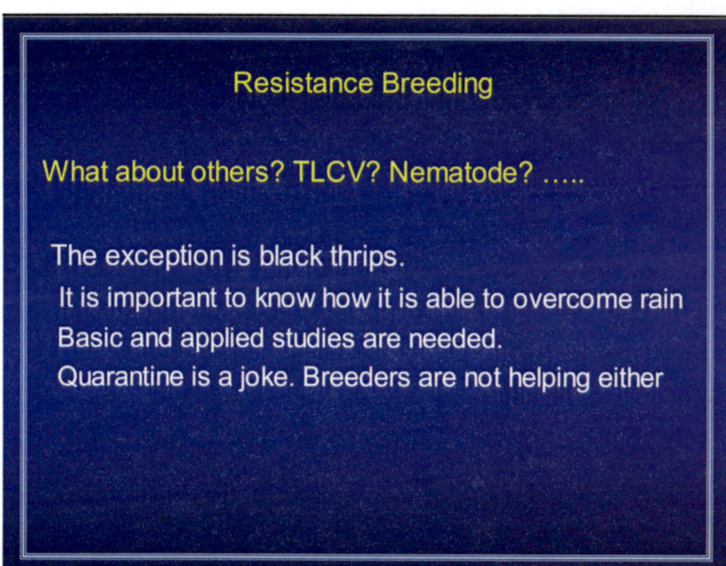

The exception for this, of course, is black thrips. It still requires some basic work to be done. I would request some entomologists, in association with others to understand how black thrips overcome rain. Once that is understood, I think, we must attempt to do breeding for resistance to black thrips. There is a lot of basic work that is needed at this point of time and there is no silver bullet as an answer for black thrips control. The second one is that I would also request to focus on strengthening quarantine systems and their functioning. Kadayam in Andhra Pradesh has become an epicentre for floriculture trade. There are also lot of testing introduced material that is openly evaluated without proper evaluations, and nobody is reporting any problem. For example, we imported some grape genotypes and they were quarantined at the NRC for Grapes. All of them turned out to be infected and the entire introduced material had to be destroyed. We should be cautious while introducing materials.

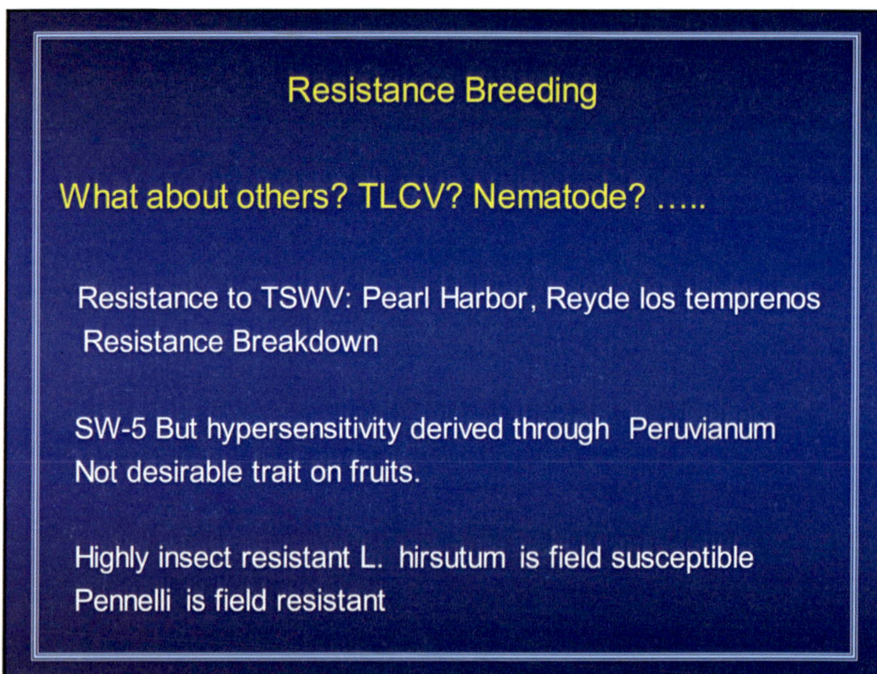

Pearl Harbor was one of the first varieties that was developed in Hawaii for resistance to tomato spotted wilt virus. Then, there was also *Rey de los temprenos* variety developed in Latin America for TSWV resistance. However, the resistance soon broke down. Many may be thinking that SW5 is hypersensitive which is derived from *L. peruvianam*. This does not mean that the other derivatives from *L. peruvianam* should not be really looked into. *L. pinnellii* is field resistant should be relooked into.

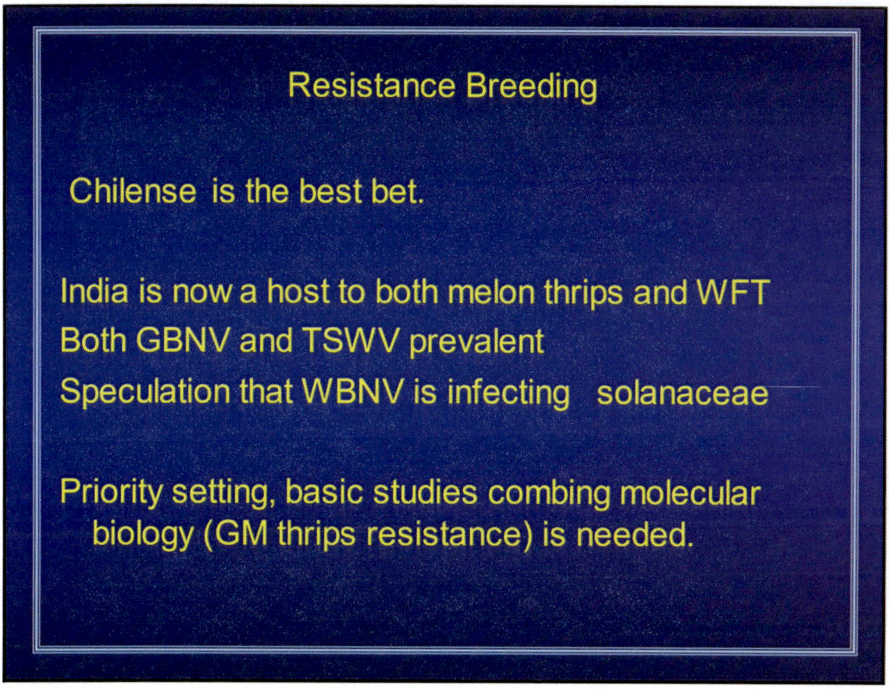

L. chilense is the best. India is now a host to both melon thrips and Western Flower Thrips. Western countries usually have Western Flower Thrips while many other countries do not have the kind of the crops that we have. The dependency of the crop that we have and the type of very efficient viruliferous thrips that we have calls for serious attention. We have both GBNV and TSWV. My presumption is that TSWV is going to overtake GBNV in the days to come. Research priority setting should include basic studies on thrips with molecular biology. In the United States of America, genetically modified thrips resistance in cotton is developed. Maybe, we should look into what kind of mechanisms are there.

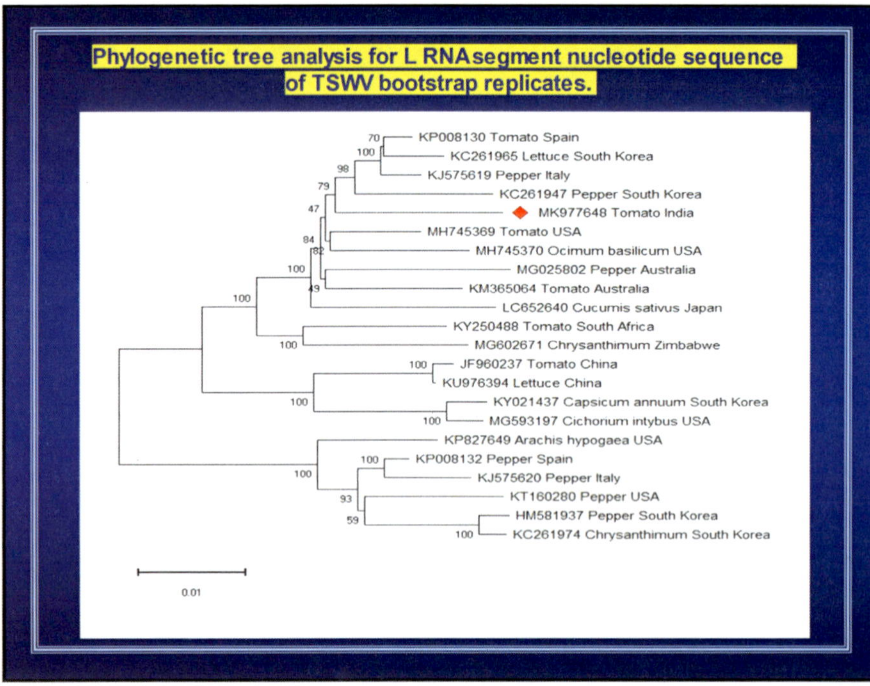
Phylogenetic tree analysis for L RNAsegment nucleotide sequence of TSWV bootstrap replicates.
KP008130 Tomato Spain
KC261965 Lettuce South Korea
KJ575619 Pepper Italy
KC261947 Pepper South Korea
MK977648 Tomato India
MH745369 Tomato USA
MH745370 Ocimum basilicum USA
MG025802 Pepper Australia
KM365064 Tomato Australia
LC652640 Cucumis sativus Japan
KY250488 Tomato South Africa
MG602671 Chrysanthimum Zimbabwe
JF960237 Tomato China
KU976394 Lettuce China
KY021437 Capsicum annuum South Korea
MG593197 Cichorium intybus USA
KP827649 Arachis hypogaea USA
KP008132 Pepper Spain
KJ575620 Pepper Italy
KT160280 Pepper USA
HM581937 Pepper South Korea
KC261974 Chrysanthimum South Korea
0.01

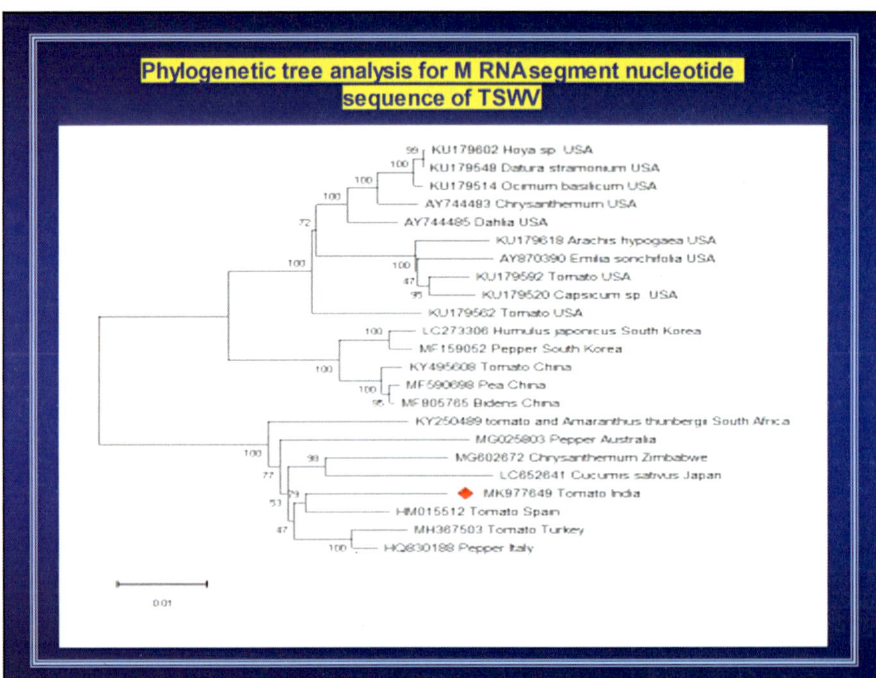
Phylogenetic tree analysis for M RNAsegment nucleotide sequence of TSWV
KU179602 Hoya sp. USA
KU179548 Datura stramonium USA
KU179514 Ocimum basilicum USA
AY744483 Chrysanthemum USA
AY744485 Dahlia USA
KU179618 Arachis hypogaea USA
AY870390 Emilia sonchifolia USA
KU179592 Tomato USA
KU179520 Capsicum sp. USA
KU179562 Tomato USA
LC273306 Humulus japonicus South Korea
MF159052 Pepper South Korea
KY495608 Tomato China
MF590698 Pea China
MF805765 Bidens China
KY250489 tomato and Amaranthus thunbergii South Africa
MG025803 Pepper Australia
MG602672 Chrysanthemum Zimbabwe
LC652641 Cucumis sativus Japan
MK977649 Tomato India
HM015512 Tomato Spain
MH367503 Tomato Turkey
HQ830188 Pepper Italy
0.01

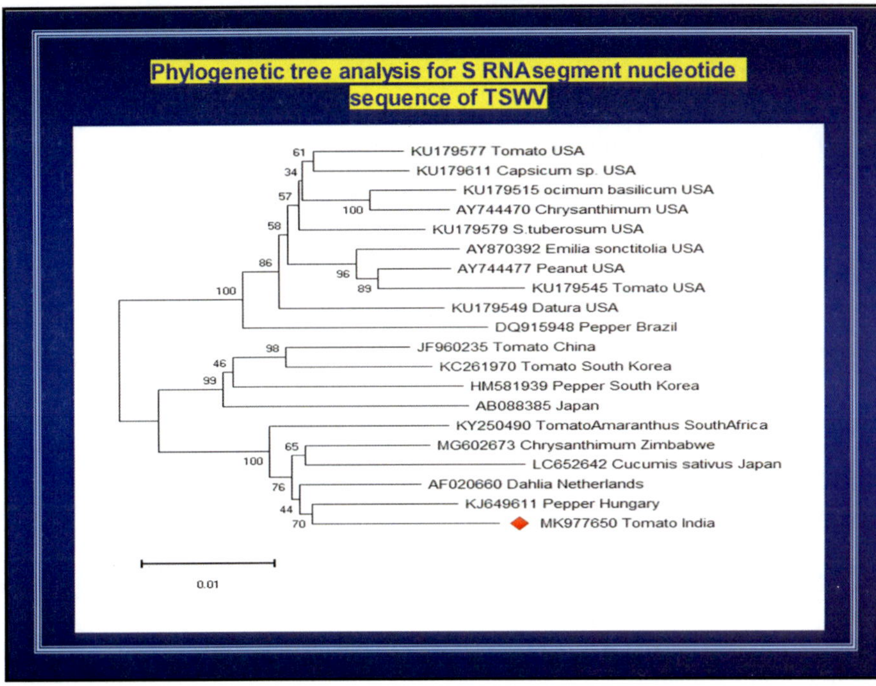

Already work is being carried out on the phenotypical phylogenetic tree of the L-RNA, M-RNA and S-RNA of TSWV and this analysis should help us to understand the tospo viruses better.

Questions and Answers

Kiran: I am Kiran working as a researcher for the last 10 plus years in an Indian private company. I am an entomologist and we are working on thrips resistance. My question is, like in case of tomatoes, where trichomes has helped thrips resistance, will it help thrips resistance in hot pepper too?

Krishna Kumar: The answer to your question is, generally, that trichomes do confer resistance across a range of insects. However, what we should be looking for is whether it is glandular or non-glandular trichomes. Second is that in tomato, better to focus on tospo virus resistance than thrips resistance.

Suren Tikoo (Moderator): His question was whether you would suggest any species or any accession in hot pepper that they can consider as source for thrips resistance which is due to trichomes.

Krishna Kumar: I would leave it to the breeders because they will know which is the species to use in breeding. There is a simple assay called the 'Thumbtack Assay' which can be used for this purpose. Leave a small thrips nymph on a thumb tack pinned to a leaf (different genotypes of chilli or wild species). If there are trichomes, the thrips nymphs take maximum time to descend on the leaf. It is a very simple method by which we can know whether the trichomes confer resistance or not depending on the time taken to descend. It is also useful in screening resistance to mites.

Harish: I am Harish. I would like to know the number of thrips to be used and also the stage of chilli seedlings to be released in artificial screening for thrips tolerance.

Krishna Kumar: In case of chillies, I would request you to please see our publication in Euphytica which was published in 1996 or so, where extensive screening of a nearly 200 genotypes was carried out. Screening of early-stage seedlings is highly correlated with the susceptibly at the later stages of the plants. So, we can go for seedling screening with regard to thrips. With regard to lab vs field screening, I would say that instead of going directly to laboratory screening, the first step should be to carry out field screening and then those genotypes which are apparently resistant should then be subjected to screening with a population of *Scirtothrips* or black thrips. I would suggest you to do field screening first which is much more reliable. If you have a larger number of germplasm lines, I suggest, that only the ones promising in the field screening to be further tested in the laboratory.

Question (from chat box): What are the control measures against black thrips?

Krishna Kumar: There are only two chemicals for black thrips control - one is Electro and one more. The best of the pesticides promises 60 per cent control, while a realistic field control may be 40 per cent. There is nothing which is really working at this point of time against black thrips to the level of desired satisfaction.

Pradeep Kumar: Can glandular trichomes give control to all thrips diseases?

Krishna Kumar: It is a very interesting question. In case of thrips, what are we looking for – is it resistance to thrips or resistance to thrips-transmitted virus. If we are looking for thrips resistance, the space that is required for nymphs and the space that is required for adults are different. If we have high density of trichomes, then the nymphs also can be trapped. The proportion of the trichomes versus glandular trichomes varies. Is the proportion of trichomes is less than the glandular trichome? What the glandular trichomes do is the small bulb like structure that contains the exudate at the tip of the trichome, gets attached to the hind legs of the thrips, which, all its life, will be only trying to remove it and die. But in case of adults, they require more space and they don't colonize. But if they come for egg laying like in case of chillies, it is good. I would suggest to initially look for a greater number of glandular trichomes and higher density of trichomes. These two factors contribute to our understanding of trichomes. But in tomato and other crops, I would say that the information is very meagre. There is work going on with regard to acyl sugars and whitefly resistance in other universities in the US for white fly resistance. I would request you to look into that.

Suren Tikoo (Moderator): Cornell University has been working on acyl sugars for the last 30 years. Finally, they have traced it to chromosome 3, but there are too many linkage drags to use it in breeding. I would think that it is a long-term investment and we need to keep working on it.

Question (From Chat Box): Is there any bio control agent to control thrips?

Krishna Kumar: There are a number of biocontrol agents that are available to control thrips. For example, there is a predator, *Orius*, which is a very good predator, which is usually found in numbers that are not sufficient to keep the pest density of thrips to a desirable level. I would also add that if we take up one spray, the predators would get eliminated too. There is currently a slow shift going on from the predators to the fungus, *Bavaria*, and many others. If we can isolate and use them as bio control agents, it will help black thrips control, as these thrips are prevalent in high humidity region. This approach should really be looked into. There is another, a nematode called *Thripanema,* which feeds on thrips like entomopathogenic nematodes. While working on the molecular systematics of thrips, we hit upon *Tripanema,* but unfortunately the nematode was dead. We are still trying to collect *Tripanema* from India. So far, we have not been very successful but it is available in US and other places. The advantage is that we can deliver this EPN along with drip irrigation and get them to the root zone of the plant. In protected cultivation it can be an excellent option for bio control of the pseudo-pupal stages of the thrips. So, there are bio control methods. Parasitoides are very few and not very effective.

Suren Tikoo (Moderator's Summary): Thank you Dr. Krishna Kumar for the exhaustive presentation on thrips. I would summarize the three key points made by Dr. Kumar:

- It is clear from the presentation that thrips is one of the major problems in many of our vegetable crops. There is no effective insecticide at the moment which is giving 100 per cent control.
- We should not only have genetic resistance to the thrips, but also need to have resistance to the virus as well as resistance to the insect vector that is transmitting, whether it is to thrips or TLCV or any other insect-transmitted virus. So having both would probably enhance the resistance.
- The work that is going on in utilization of crop wild relatives in breeding

is not enough. I personally feel that the public sector researchers should spend more time on this pre-breeding work. We should come up with public-private partnerships consortia for each of these viruses. The results would be quicker and gains would be wider. We need more and more research on crop wild relatives because today whatever we have in terms of resistances they have come from the crop wild species. *L. peruvianum* has been very useful for tospo virus resistance. Especially we have found GBNV resistance in the two *L. peruvianiam* collections that came from AVRDC.